T0281174

Life and Death Rays

Life and Death Rays
Radioactive Poisoning and Radiation Exposure

Alan Perkins

CRC Press
Taylor & Francis Group
Boca Raton London New York

CRC Press is an imprint of the
Taylor & Francis Group, an **informa** business

First edition published 2021
by CRC Press
2 Park Square, Milton Park, Abingdon, Oxon, OX14 4RN

and by CRC Press
6000 Broken Sound Parkway NW, Suite 300, Boca Raton, FL 33487–2742

CRC Press is an imprint of Informa UK Limited

British Library Cataloguing-in-Publication Data
A catalogue record for this book is available from the British Library

Library of Congress Cataloging-in-Publication Data
A catalog record for this book has been requested

ISBN: 978-0-367-46336-6 (hbk)
ISBN: 978-0-367-45649-8 (pbk)
ISBN: 978-1-003-02824-6 (ebk)

Typeset in Times
by Apex CoVantage, LLC

For Pauline, Heather and Frances
Thank you for everything.

Contents

Preface

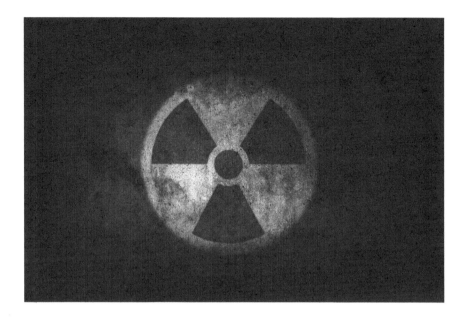

When most people think of radiation and radioactivity, they think of the atomic bombs dropped on Hiroshima and Nagasaki, the Chernobyl and Fukushima disasters, the fear of getting cancer or perhaps gaining superpowers. After all, it is what changed Peter Parker into Spiderman and Bruce Banner into the Incredible Hulk! It can be scary stuff, but radioactivity is natural and is all around us in our homes, in the ground beneath us, in the sky above us and in the food we eat. Since the discovery of radioactivity over 120 years ago, it has been used for a range of purposes, in nuclear weapons, power generation, industrial processes, medical diagnosis and treatment, and as a poisonous murder weapon. At one time people thought a dose of radiation would do you good. You could buy radium tonics and atomic soda and take friends to have an X-ray to see what they looked like inside. Over time strange things began to happen and a lot of people died from the horrible effects of high radiation exposure, but the really good thing was that radiation could also be used to diagnose and cure cancer as well. This weird irony all comes down to the difference between a poison and a medicine.

I have been fortunate to have had a fascinating and rewarding career working as a Clinical Scientist in Medical Physics and Nuclear Medicine at the Nottingham University Hospitals and as a Clinical Academic in the School of Medicine at the University of Nottingham. Over that time, I never stopped learning about the hazards and beneficial applications of radiation and now and then unexpected events such as the Chernobyl disaster, the London polonium poisoning and the medical radionuclide crisis took me away from the more normal aspects of clinical scientific work. This book gives a short history of the discovery and uses of radiation and radioactivity which have literally changed the course of history. It contains a collection of stories and factual details that I have collected over the years and used in conference presentations and published articles. In an attempt to make the subject more widely accessible and to broaden the readership, I have attempted to describe the events with the minimal use of scientific terms. However, to help understand the subject I have had to include some basic terms covering radiation, radioactivity and radiation dosimetry. Descriptions of the basic units of activity, radiation doses and biological effects are briefly described in Chapter 3 and are given in more detail in Chapter 11.

I have deliberately concentrated on the human aspects of the discovery, uses and effects of radiation and radioactivity and described the good, bad and downright ugly consequences that followed. Humanity has had to understand and respect radiation and to learn how to use it safely, whether in power generation, industry or medicine. These are stories of fascinating science, tragic misadventure and malicious intent.

Alan Perkins
Emeritus Professor of Medical Physics
Radiological Sciences
School of Medicine
The University of Nottingham

Acknowledgements

I am extremely grateful for the help, encouragement and advice received from my close friend and colleague John Lees, Emeritus Professor in Bioimaging at the Space Research Centre at The University of Leicester. His careful scrutiny of the text and timely comments were invaluable for helping me complete the project during the 2020 lockdown. I am also indebted to my colleagues in Medical Physics at Queen's Medical Centre in Nottingham, especially Dr David Pye for his advice on the radiation physics and radiation protection sections of Chapter 11. Thanks also to Pauline Perkins and Heather Perkins for careful proof reading and helpful comments with the manuscript.

About the Author

Alan Perkins is Emeritus Professor of Medical Physics in Radiological Sciences in the School of Medicine at the University of Nottingham, having previously held the position of Clinical Professor of Medical Physics in the School of Medicine and Honorary Consultant Clinical Scientist at Nottingham University Hospitals NHS Trust, where he was the Clinical Support Divisional Lead for Research and Innovation. He has over 35 years of experience in Nuclear Medicine and Medical Physics and broad managerial experience in the National  Health Service. His contribution to clinical science and collaborative research has resulted in authorship of over 200 peer-reviewed publications and 6 published textbooks.

Professor Perkins is Past President of the British Nuclear Medicine Society and Past President of the International Research Group for Immunoscintigraphy and Therapy, previous Vice President of the Institute of Physics and Engineering in Medicine and is currently a Governor and Chair of the Research Strategy Board for Coeliac UK. He is an Honorary Fellow of the Royal College of Physicians of London, a past editor of the UK journal *Nuclear Medicine Communications* and for over 9 years represented the UK on The High-Level Group for the Security of Medical Radioisotope Supplies at the OECD in Paris. He has consulted for a number of commercial organisations and has acted as an expert witness for pharmaceutical litigation in the US.

1 ☢

The Poisoned Chalice

A poison in a small dose is a medicine
and a medicine in a large dose is a poison.
Alfred Swaine Taylor, 19th-century toxicologist.

1.1 POISONING OF A RUSSIAN DEFECTOR

A Russian defector lay in a hospital bed. He had been admitted following a period of sickness culminating in violent episodes of vomiting. These vomiting attacks were so severe, that at times he was losing consciousness. The doctors initially diagnosed acute gastritis and instructed the patient to rest. For 5 days he was treated for gastritis, with minimal medication, and no detailed investigations were ordered. His symptoms persisted, and a visitor noticed a change in his skin colour to a deep copper red, which the attending doctors had initially assumed to be his natural complexion. On the sixth day after a restless, uncomfortable night, his symptoms changed. He awoke to see his pillow stained with blood. On entering the room, the nurse was astonished to see his face covered in black and blue swellings. She left abruptly to seek medical assistance while the patient looked into the mirror to see his face, dry and drawn with dotted dark spots, oozing blood. His eyelids were swollen and bursting with sticky secretions. He reached up to his head, examining a tuft of hair which he pressed down on his scalp. Moving his arm to his side, the hair remained in the palm of his hand. Reaching up again, he pulled at his scalp, clutching lock after lock of hair which fell around him and on to the floor.

After an examination by the chief physician, it was considered that the patient has experienced a rare allergic reaction which was treated with injections of penicillin, calcium and vitamin supplements. The doctors considered their patient with some curiosity. The professorial medical team were summoned to undertake further examinations and went through the list of drugs that the patient had been given since he had been admitted to hospital. His hair loss was a particular concern. The professor examined his scalp and then his chest hair, which came away by the fistful. Exasperated, the professor raised his voice and asked, 'Tell me, have you taken anything?', 'What are you suggesting?' the patient replied. 'Did you try and kill yourself?' the professor replied in anger.

'I did not do this to myself. I have not tried to commit suicide, but there are people in this world who would consider it advantageous if I *was* poisoned, for I am a Russian and I have recently left the Soviet Union. There are former bosses who would like to square accounts with me. Poisoning would not be beyond them. Such things have happened there before!' The professor stood up, 'In that case, further measures must be taken and it is my duty to inform the police. It may be that the food or whatever substance was given to you has been served to somebody else'.

The remote prospect of poisoning had not previously been considered, but the medical team now had to take this seriously and undertake a complete reassessment of the patient's condition. A series of medical specialists was brought in to examine the patient, and further blood samples were taken for laboratory analysis. However, on examination of the lab results the team remained confused. Evidence of the more commonly used poisons such as arsenic, strychnine and hydrocyanic acid were absent, and the medics surmised that the only conceivable solution to the patient's condition was that he has been given a combination of poisons, which included the metal thallium. This had not only turned into a complex medical case, it had become a criminal investigation with international political implications. Furthermore, contrary to what the reader may assume, these events did not take place in London in 2006 and the patient was not Alexander Litvinenko. He was Nikolai Evgenievich Khokhlov, a Russian defector poisoned with thallium by foreign agents in Frankfurt in 1957. At that time deliberate radiation poisoning was not even a remote consideration. In reality, as the patient deteriorated, worse was yet to come.

1.2 POISONS AND POISONERS

A poison can be described as 'any substance which when introduced into or absorbed by a living organism destroys life or injures health'. Ancient man initially used plant substances such as curare and aconite to poison their hunting arrows and darts. It soon became apparent that what could bring down a beast would also kill a man. The word 'toxin' is derived from the Greek word *toxikon*, which comes from the word 'poison' which had a particular meaning related to the poison in which the arrows were dipped. The word 'intoxicated', which is a Latin derivation, was the sickness resulting from poison or poisoning. Toxicology, or the science of poisons, is the study of the adverse effects of chemicals or physical agents on living organisms. Tissue and organ damage may take many forms, from immediate death to subtle changes not apparent until months or years later [Royal Society of Chemistry, 2015]. Paracelsus (1493–1541), commonly known as the 'Father of Toxicology', determined that a plant or animal poison contained specific chemicals that were responsible for toxicity. Crucially he documented that the body's response to those chemicals depended on the amount of the substance (the dose) received. Paracelsus is often quoted for his statement 'All substances are poisonous; there is none which is not a poison. The right dose differentiates a poison and a remedy'. His work laid the foundation for transforming medical science from medieval to modern practice. He believed that certain substances, such as arsenic, mercury and lead, could be beneficial in the treatment of disease, if administered in very small controlled doses. The route of exposure is important in determining toxicity. Some substances may be highly toxic by one route but not by others, for example many snake venoms are highly toxic if they enter through the bloodstream, but they may be harmless when swallowed. Two major reasons for this are differences in absorption and distribution in the body. For example, ingested chemicals when absorbed from the intestine may be degraded by the acid environment of the stomach and are first distributed via the hepatic portal vein to the liver and may be detoxified by metabolism. The metabolites themselves may be more toxic, less toxic or of equivalent toxicity to the ingested parent chemical.

Orfila was a Spanish physician who lived from 1787 to 1853 and established the founding principles of modern forensic toxicology. He identified the correlation between the chemical and biological properties of poisons of his time and demonstrated the effects of poisons on specific tissues by analysing autopsy materials for toxins and assessing the associated tissue damage. He is best known in French Legal Medicine for demonstrating the presence of arsenic in tissues using the 'Marsh test' which had been developed by James Marsh, a British chemist in 1836.

1.3 ENVIRONMENTAL AND INADVERTENT POISONING

Many natural substances are hazardous, especially if they occur in concentrated amounts. In the 18th and 19th centuries mercury was used in the production of felt, a material commonly used in hat making. Workers in the hat making industry of Luton in Bedfordshire, England, developed dementia-like symptoms due to mercury poisoning, giving rise to the colloquial term 'Mad as a Hatter'. Hat makers in Danbury, Connecticut, US, were later observed to have similar symptoms including slurred speech, tremors, stumbling and even hallucinations, a condition known locally as the 'Danbury Shakes'.

A more pervasive example of long-term poisoning has come from the use of lead, a heavy dense metal that is easily extracted from ore by crushing and heating to form a pliable metal that is easy to shape into sheets or pipes. The Romans used lead to make water pipes 2,000 years ago, and even at that time there were concerns about the effect of lead on health. Lead mining has continued over centuries despite the miners often ending up either insane or prematurely dead. The Peak District in Derbyshire, UK, is scattered with old mine sites including one of the oldest known lead mines in Britain. The local Derbyshire saying captures the essence of those mining and extracting lead,

> *Derbyshire born and Derbyshire bred*
> *Strong in the arm and weak in the head.*

For many years lead was added to paint to quicken drying and increase durability. In 1678 workers who made lead white pigment for paint were described as suffering ailments including dizziness in the head, with continuous great pain in the brows, causing blindness and stupidity. Adding lead to paint was banned in the US in 1978, and the sale of lead paint was banned in the EU in 2003. However, the paint industry was not alone in the use of lead as an industrial additive.

In retrospect, it seems insane to take a known toxic element, put it into machines that pump it out in breathable form and then move these machines millions of miles a day all over the world, along roads of cities, towns and villages, so that billions of people would constantly inhale the poison. But in August 1924 General Motors and its part-owner DuPont joined with the Standard Oil Company and the Ethyl Corporation to market leaded gasoline. It was claimed that adding tetraethyl lead to fuel allowed car engines to run more smoothly and quietly. Other additives would have performed the function just as well, but since this was patented and would make money it was aggressively promoted. Standard Oil was aware of the problems of producing ethyl lead in large quantities. The laboratory where it was developed was known as 'the loony gas building'. The first production line in Ohio was shut down after the death of two workers. There were also fatalities at the New Jersey laboratory, where workers had repeated hallucinations of insects, causing the laboratory to be known as 'the house of butterflies'.

Lead in petrol was used for over three-quarters of a century until it was banned as a result of overwhelming adverse medical evidence. Low levels of lead have been linked to high blood pressure, stroke, heart and kidney disease and accelerated ageing. It is widely considered that children's brains are especially susceptible to chronic lead poisoning and the effects proved worse if you happened to live near a major road or motorway. Exposure to lead at a young age can result in learning disabilities, lower IQ and attention-deficit/hyperactivity disorder. There is also a hypothesis that the removal of lead from motor fuel in the US led to a reduction of lead exposure to children and a subsequent reduction in violent crimes. This is just one story of disputed science and delayed regulation due to industrial interests and political pressure, but the same goes for asbestos, tobacco and other products that can slowly kill. The industrial process of adding lead to petrol was probably the greatest ever mass human poisoning experiment that has ever occurred.

With increasing industrialisation, the utilisation of ores, minerals, chemicals and radioactive elements has created increased risks to human health, through either inadvertent or intentional use. A list of some of the main poisoning incidents involving man-made substances is shown in Table 1.1.

TABLE 1.1
List of major man-made poisoning incidents

Date	Incident	Recorded deaths
1857	Hong Kong, Esing Bakery Incident: 300–500 people consumed bread adulterated with large quantities of arsenic. Only three deaths were recorded, since the amount of arsenic was only high enough to induce vomiting and prevent digestion. It is unknown whether the contamination was deliberate or accidental	3
1858	England, Bradford sweets poisoning: Sweets accidentally made with arsenic were sold from a market stall which led to the poisoning of more than 200 people, resulting in 21 deaths	21
1971	Iraq, Poison grain disaster: A mass poisoning by grain treated with a methyl mercury fungicide which was imported to the country as seed and never intended for human consumption. According to several estimates, the recorded death toll varies from 459 to 650 people, though much higher estimates have been offered	459–650
1981	Spain: An outbreak of toxic oil syndrome supposedly caused by contaminated colza oil killed over 600 people	600
1984	India, The Bhopal tragedy: A gas leak in the Union Carbide pesticide factory led to at least 3,787 deaths	3,787

Date	Incident	Recorded deaths
1988	England, Camelford water pollution: The accidental contamination of the drinking water supply to 20,000 local people and up to 10,000 tourists with 20 tonnes of aluminium sulphate. Officially, there were no deaths caused by the accident	0
1994	Japan, Matsumoto Sarin attack: It was originally aimed at three judges who were overseeing a lawsuit concerning a real-estate dispute perpetrated by members of the Aum Shinrikyo doomsday cult in Japan's Nagano prefecture. On the night of June 27, aerosol was released from a converted refrigeration truck in the Kaichi Heights. Eight people were killed and over 500 were harmed	8
1995	Japan, Tokyo subway sarin attack: It was an act of domestic terrorism perpetrated on 20 March 1995 by members of the Aum Shinrikyo doomsday cult movement. In five coordinated attacks, the perpetrators released sarin on three lines of the Tokyo subway during rush hour, injuring over 1,000 people	13
2010	Nigeria, Zamfara lead poisoning: It is thought that the poisonings were caused by the illegal extraction of lead ore by villagers, who took crushed rock home to extract the lead. This resulted in the soil being contaminated which then poisoned people through hand-to-mouth contamination	163
2016	Pakistan, Punjab sweet poisoning: An accidental contamination of baked confectionery with illegal agriculture pesticides led to at least 33 deaths	33

1.4 DELIBERATE POISONING

No other form of murder has been documented, dramaticised, romanticised or has achieved such notoriety as poisoning. Throughout history, poisoning has been a favourable murder weapon. The Greeks, Egyptians, Romans and Persians all documented various forms of poisoning. In particular, the Egyptians were notable as the first masters of distillation of many classical poisons, including the extraction of poison (probably cyanide) from peach kernels. By 82 BC, poisoning had apparently become so widely used in the Roman Empire that the Roman dictator and constitutional reformer Lucius Cornelius Sulla found it necessary to issue the world's first law against poisoning, called the *Lex Cornelia*, which in practice had little effect.

By the Middle Ages, poisons were regularly sold to the general public by apothecaries. Although religion was responsible for the demise of scientific study and knowledge in many fields in the West, knowledge of poisons continued to increase. Many academic texts on poisons were written by monks, for example, *The Book of Venoms* written in 1424 by Magister Santes de Ardoynis, which described the known poisons at the time, how they worked and how they could be given. During the 14th and 15th centuries, Italian alchemists studied and experimented with substances to increase the potency of poisons derived from classical sources. This information spread from Italy to Paris, thanks to the efforts of Queen Catherine De Medici, and led to a boom in the poisoners' industry, as poisoning occurrences spread across Europe. By 1572 at least 30,000 'sorcerers' were undertaking poisoning activities in the streets of Paris. In the 16th century a poisoner-assassins' guild called the Council of Ten was established by a group of alchemists in Italy to provide 'elimination' services for a fee.

In 1589 a publication called *Neopoliani Magioe Naturalis* written by Giovanni Battista Porta served as a textbook for poisoners. This described how wine could be laced with a deadly concoction called *Venenum Lupinum*, which was made from the aconite plant, common yew, caustic lime, arsenic, bitter almonds, powdered glass and honey and shaped into walnut-sized pills. By the 17th century, schools of poisoning had been established in Venice and Rome, where individuals could advance their poisoning intentions by forming secret societies in which they received instructions on the nature and means of administering poison. Members of the nobility became particularly anxious, as many of them were targets of poisoning. Attempts were made on the life of Queen Elizabeth I, including an unsuccessful attempt by a Jewish physician called Dr Lopus, who was hanged, drawn and quartered for his crime.

In the 19th century during the reign of Queen Victoria, poisoners had the means and increased motivation for committing numerous murders by poisoning. Poison was readily available to the general public in the new urban areas and industrial cities for the eradication of pests and infestations. The introduction of life insurance policies gave an added incentive for personal gain from the untimely demise of someone else.

Poisoning became so widespread in the UK that laws such as the Arsenic Act of 1851 were introduced in an attempt to bring this particular crime under control. Over the years that followed, new sophisticated poisons emerged.

In the 20th century, published literature concerning the manufacture of poisons was widely available and deadly chemicals were being produced on an industrial scale, albeit for legitimate purposes. The average household medicine cabinet or store cupboard was a would-be poisoner's treasure chest, with thallium, radioactive radium and morphine included in everyday products such as medications, pesticides, paints and cosmetics. Although some poisonings occur as a result of ignorance and inadvertent ingestion, the act of deliberate poisoning is still regarded as a particularly intriguing crime. In the US in 1929 the Wall Street Stock Market crashed, sending the US economy into a depression. By the end of 1930 over four million Americans were unemployed. Murder by poison became an increasingly common crime. In 1953 the first US poison control centre was opened in Chicago. By 1957 there were 17 poison control centres in the US and forensic investigation had become an important specialty of criminal science.

1.5 THALLIUM: THE POISONER'S POISON

The discovery of thallium was controversial. William Crookes of the Royal College of Science in London published the discovery of thallium in *Chemical News* in March 1861, after observing a prominent green flare in the spectrum of a sample of impure sulphuric acid. Although he had discovered a new element, he did little research into its properties, being more interested in the new electrical devices of the time. Shortly after, in Lille, France, Claude-August Lamy began more detailed research into thallium and separated enough of the metal to cast a small ingot, which he sent to the London International Exhibition of 1862. The French Academy credited him for the discovery of this new metallic element and he was awarded the Exhibition Medal. Crookes was furious, having previously published his findings, so the exhibition committee awarded him a medal as well.

Thallium is a heavy chemical element with the symbol 'Tl' and atomic number 81. There are 25 isotopes of thallium with atomic masses ranging from 184 to 210. The only stable isotopes of the metal are thallium-210 and thallium-205, the rest being radioactive. It is a soft, malleable, grey metal that is not found freely in nature. The name thallium comes from the Greek word *thallos,* which means green shoot or twig. This name was given to the metal as a result of the green flame it produced when heated in a flame test, as in William Crookes' initial observations. When separated in a pure form, thallium resembles tin, but it becomes discoloured when exposed to air.

Modern-day use is mainly restricted for industrial and research purposes, and one particular radioactive form of thallium (thallium-201 thallous chloride) is used in medical diagnostics (see Chapter 8). The current uses of thallium include electronics manufacturing (thallium sulphide and thallium selenide are used in electronic devices, switches and closures, primarily for the semiconductor industry), optical glass manufacturing (thallium bromide and iodide are used in infrared optical materials and scintillation radiation detectors) and high-temperature superconducting materials, low-temperature thermometers and green fireworks.

Pure thallium and compounds containing thallium are highly toxic to animals and humans. It is particularly dangerous to humans because thallium and its compounds readily dissolve in water and have no taste. It can also be readily absorbed through the skin. The biological action and medical symptoms of thallium poisoning are distinct but can vary depending on the nature, severity and timing of the exposure. Thallium poisoning can present either as an acute poisoning or as a long-term chronic poisoning. The initial symptoms of thallium poisoning are similar to those of a common cold or flu and also include stomach cramps and diarrhoea. If large amounts of thallium have been administered over a short time period, individuals will show signs and symptoms of acute poisoning with the classic symptoms of severe stomach pain, nausea, vomiting and diarrhoea within 3–4 hours of exposure. At lower doses, the symptoms can develop within 2 to 5 days of exposure. Once in the body, it acts on enzymes in brain and muscle and can produce numbness and rapidly progressing peripheral neuropathy with pain and tingling, especially in the palms of the hands and soles of the feet 2–5 days after exposure.

Internally thallium exhibits some similarities with the body's essential alkali metal cations, particularly sodium and potassium (due to their similar ionic structures) which are essential for the regulation of fluid balance, muscle contraction and nerve pathways. The metabolism of thallium via the potassium uptake pathways disrupts many cellular processes by interfering with the function of proteins that incorporate cysteine, an amino acid containing sulphur. Ultimately it attacks the central nervous system, resulting in coma and blindness. Sudden and dramatic hair loss that progresses to widespread alopecia about 2–3 weeks after exposure is one of the characteristic dermatological signs of thallium poisoning. Hair loss due to atrophy of the hair follicles primarily affects the scalp, temporal parts of the eyebrows, the eyelashes and the limbs. Hair discoloration may also occur. Skin rash, redness and scaling of the palms and soles and pustular eruptions of the face can occur. Transverse white lines on the nails (Mee's lines) can appear about 1 month after the poisoning.

Chronic thallium poisoning can occur after months or years of repetitive exposure to small amounts. Thallium can be absorbed through the skin, respiratory and gastrointestinal tract accumulating to toxic levels. Because the presentation of chronic thallium poisoning appears similar to other diseases, many cases of industrial thallium exposure may go undetected. Signs and symptoms of chronic poisoning include tiredness, headaches, depression, hallucinations, psychosis, dementia, poor appetite, leg pains, hair loss and disturbances of vision. Hair roots may have dark brown or black pigmentation. With chronic exposure, these darker regions appear in bands, demonstrating multiple thallium exposures. In cases of both acute and chronic thallium poisoning death finally occurs due to either respiratory paralysis or circulatory disruption.

From the year 1700 to 1900, thallium was used as a rather dubious treatment for ringworm, a fungal infection of the skin, which resulted in an itchy round rash. Although it could eradicate the ringworm, it also had dire consequences for the patients. In the past thallium salts have also been used in the treatment of syphilis, gonorrhoea and tuberculosis. The toxic nature of thallium resulted in its widespread use as an insecticide and rat poison. In Guyana in the 1920s, the use of thallium sulphate as a rat pesticide had unintended consequences. It was spread around areas of rat infestation and became absorbed into plants, which were eaten as vegetables and through the food chain in meat, resulting in 40 recorded deaths. Surprisingly, the use of thallium for this purpose in Guyana continued until the 1980s when blood and urine testing in Georgetown hospital revealed that 77% of 2,400 patients tested positive for evidence of thallium poisoning.

The fact that thallium is colourless and odourless and has no taste meant that it became notorious as a means for murder. It was previously described as 'the poisoner's poison' or 'inheritance powder' because of its ready availability and undiscernible nature when mixed with food and drink. Thallium sulphate (Tl_2SO_4) produced by dissolving thallium in sulphuric acid forms a white powder with similar appearance to that of granulated sugar. The lethal dose for humans is 15–20 milligram per kilogram. Ingestion of between 500 milligram and 1 gram would result in death. In other words, one teaspoon would be enough to kill 34 people. Conveniently, the delayed onset of severe symptoms other than gastric and flu-like symptoms allows the poisoner to be long gone before suspicions are aroused.

An early notable perpetrator of thallium poisoning was Graham Young, who was born in Neasden, London, in 1947. Obsessed with poisons from an early age, Young would test various toxic substances on his relatives. Members of his family became seriously ill, including his sister who was taken to hospital after hallucinating on a train journey to work. He eventually killed his stepmother with poison, convincing his father to have her cremated to destroy the evidence. When his sister became ill again, his aunt became suspicious and took him to a psychiatrist, who subsequently alerted the police. Young was sent to Broadmoor Hospital and eventually released in 1971 when the authorities believed he has been cured. Following his release, he found work as a laboratory technician at a chemical factory in Bovington near Hemel Hempstead. Soon after he started his new job, a mysterious illness in the local area, known as the 'Bovingdon Bug', killed three people and affected over 70 more. Young was poisoning his workmates' tea with thallium sulphate.

In Australia in the early 1950s, there were a remarkable number of murders and attempted murders by thallium poisoning, a series of incidents later referred to as the 'Thallium Craze'. Thallium sulphate was sold over the counter in New South Wales as a commercial rat bait, under the brand 'Thall-rat' for the eradication of chronic rat infestations in Sydney and other overcrowded inner-city suburbs. Yvonne Fletcher was convicted of the murder of her two husbands. She originally got away with murdering the first husband, but when her second husband exhibited the same symptoms, friends became suspicious. Other cases include Beryl Hage, who, in Sydney in 1953, confessed to putting Thall-rat in her husband's tea. At the trial she said she 'wanted to give him a headache to repay for the many headaches he had given her' in past domestic violence. In the same year Caroline Gills poisoned three members of her family and a friend by giving them tea laced with thallium sulphate. She spent the rest of her life in gaol where she became known as 'Aunt Thally'.

Even though thallium sales have been banned in the US since 1972 thallium poisoning incidents and murders still occur to this day. In 2013 Tianle Li, a Chinese-born chemist who worked for Bristol-Myers Squibb, was convicted of poisoning her husband Xiaoye Wang rather than let him divorce her. She had access to thallium from her workplace. The doctors, not having seen thallium poisoning before, failed to recognise the symptoms. The cause was identified only when a nurse recalled the case of a student from Beijing University in China in 1995 who survived a mysterious illness following treatment. The test results were positive for thallium but attempts to save Xiaoye Wang's life ultimately failed. In December 2018 Race Remington Uto, a 28-year-old who was an electrician's mate at the Point Loma Naval Base in California, pleaded guilty to three counts of premeditated attempted murder for poisoning his wife, Brigida, with thallium three times over a 5-month span. Uto was investigated by the FBI, Naval Criminal Investigative Service and a San Diego County hazardous-materials team and sentenced to 21 years in prison. Thallium is now classed as a weapon of mass destruction by the US authorities. In 1960 the leader of the Cameroon, Félix-Roland Mormié, was assassinated with thallium by the French secret service in Geneva. It is alleged that the Central Intelligence Agency (CIA) plotted to use thallium on Fidel Castro. The project progressed as far as testing on animals but did not proceed any further. The aim was to discredit him by causing him to lose his characteristic hair and beard. It is also alleged that the South African government sought to use thallium poisoning against Nelson Mandela.

The mysterious nature of thallium and its illicit use made it a popular murder weapon in crime writing and film. Agatha Christie's crime novels involve a number of poisons including strychnine, cyanide, arsenic and belladonna (deadly nightshade). The final Hercule Poirot novel *Curtain* involves multiple poisonings with morphine, physostigmine and drugging with sleeping tablets. Indeed, Poirot himself dies in the night in an intentional act where he deliberately places his heart medication, amyl nitrate, out of his reach. Poirot is the only fictional character to have been given an obituary on the front page of the *New York Times* in 1975. In 1961 Christie published *The*

Pale Horse, a novel with a plot line based on a series of poisonings with thallium. The title of the book is taken from the Revelation of St John the Divine, chapter 6, verse 8.

> And I looked, and behold a pale horse: and his name that sat on him was Death, and Hell followed with him. . . .

During the First World War, Christie qualified as an apothecaries' assistant and worked in a dispensary in 1917. The knowledge she gained at this time enabled her to write accurate description of numerous medicines and poisons, including the symptoms displayed by the thallium poisoner's victims. This novel is notable among Christie's books as it is credited with having saved at least two lives after readers recognised the symptoms of thallium poisoning from the description given in this book. Thallium poisoning has also featured in a number of movies including *Big Nothing* (2006), *Edge of Darkness* (2010), the James Bond movie *Spectre* (2015) and TV dramas such as *House*, *NCIS* and *CSI New York*.

1.6 TREATMENT OF THALLIUM POISONING

The chemical identification of thallium is now straight forward in suspected cases and thallium poisoning can be easily recognised once suspected. Early diagnosis and treatment of thallium poisoning is key to a favourable recovery. If the poisoning has been identified at an early stage, medical intervention can provide an effective cure. The treatment of thallium poisoning includes the initial stabilisation of the patient by assessing the airways, breathing and the cardiovascular system and providing oxygen if necessary. Decontamination measures should be carried out by inducing vomiting if the thallium was ingested within the last 30 minutes. Any source of contamination should be eliminated by removing clothing and washing the skin thoroughly with soap and water. Gastric decontamination should be considered with sequestering agents such as activated charcoal or Prussian blue that can bind with the metal. Activated charcoal is made from substances containing a high carbon content such as coal, coconut shells and wood. It is a form of charcoal with one of the highest known adsorbing characteristics. It is widely used for scavenging molecules, for example, in filters and cooker hoods. If given orally it can soak up metals such as thallium. Prussian blue was one of the first synthetic pigments and has a deep blue colour. The name Prussian blue originated in the 18th century when it was used to dye the uniform coats for the Prussian army. Over the years, the pigment acquired several other 'blue' names, including Berlin, Parisian and Turnbull's blue. You only have to see Vincent van Gogh's painting Starry Night to see the intense blue colour produced by this pigment. It contains iron and cyanide in the form of ferric hexacyanoferrate (II). Surprisingly, although it contains cyanide Prussian blue is not toxic to humans. When taken orally it chelates heavy metals by ion exchange, adsorption, and mechanical trapping after releasing the iron component. Once binding has taken place, absorption from the gut is blocked and the toxic metal continues along the gastrointestinal tract to be excreted in the faeces. Prussian blue has been used extensively in the treatment of radiation casualties who have ingested metal radionuclides of thallium and caesium.

1.7 THALLIUM RADIATION POISONING

The Russian defector lying in a Frankfurt hospital bed in 1957 was Nikolai Khokhlov, who was born in 1922 in Nizhny Novgorod, the administrative capital of the Volga Central District of Russia. Four years after his birth, his parents separated. He grew up with his mother and stepfather who was a lawyer. His father, who he did not know well, encouraged him in the study of music and film, which he eagerly studied. His father held a modest position in the aviation industry but he devoted his life to the service of the party and served as a commissar in the Red Army. In 1941 during German attacks on Moscow his father was convicted of making unfavourable remarks about Joseph Stalin and was transferred to a penal battalion where he died. In the same year his stepfather volunteered to defend Moscow and was killed in action. Nikolai was a member of the People's Commissariat for Internal Affairs (the NKVD), which was responsible for law enforcement until the end of the Second World War. At the age of 19 he was trained along with three other agents to attack Nazi officers whilst performing in a vaudeville show during their expected victory celebrations in Moscow. Nikolai was chosen because of his performing and whistling abilities! With the subsequent German retreat there were no celebrations and the show never took place. However, after becoming proficient in the German language, in 1942 Nikolai went on to join a successful military squad fighting behind the enemy lines. Disguised as a Nazi officer, he parachuted into German-occupied Belarus with a unit which organised the assassination of Wilhelm Kube, the German politician and Nazi Gaultier of Belarus who became known as the *Butcher of Belarus* for his brutality against the Jewish population. The main character in the 1947 Soviet film *Podvig Razvedchika* (The Secret Agent/The Scouts Exploit, Figure 1.1), a film based on the actions of Soviet intelligence officers behind enemy lines in after the Second World War, was based on the exploits of Nicolai Khokhlov. This launched a new era of popular Soviet spy films.

In 1950 Khokhlov entered the school of Languages at Moscow University. His aim was to become a postgraduate student of linguistic science and secure demobilisation from the security services. However, he remained a highly regarded operative of the Russian intelligence services and was sent to the west on numerous spying missions. In 1954 after extensive training on surveillance, driving and target practice he was briefed on the use of new silent cigarette-case weapon which delivered high-velocity poisoned bullets triggered by an electric pulse and thought to contain potassium cyanide. The Kremlin Border Guard (KGB) were to send him to Germany to supervise two other men who were to assassinate the anti-communist Russian émigré and chairman of the National Alliance of anti-Communist Russian Solidarists Georgi Okolovic. After discussion with his wife Yana, she told him, 'If this man is killed, you will be a murderer and I cannot be the wife of a murderer'. On arrival in Frankfurt he did not carry out his mission but instead took the journey to Okolovic's flat and told him that he had been sent from Moscow on the orders of the Central Committee of the Communist Party of the Soviet Union, who had ordered his assassination. He pleaded for Okolovic's help and following his failure to carry out the assassination Khokhlov defected to the US. He toured the country speaking at meetings and gatherings on the nature of the Communist dictatorship in Russia.

FIGURE 1.1   Poster for the 1947 Soviet film *Podvig Razvedchika*.

Source: Poster of *Podvig Razvedchika*. www.alamy.com/movie-poster-secret-agent-by-boris-barnet-museum-russian-state-library-moscow-image211768801. html?pv=1&stamp=2&imageid=5F815F90–77C8–4CF7-A1E9

In retaliation, the Russian authorities arrested his wife and sentenced her to 5 years of involuntary settlement in a labour camp.

In the autumn of 1956 Khokhlov returned to Europe to support the work of Russian refugees and friends whom he had come to know through correspondence and published articles supporting freedom of the Russian motherland. On Sunday, 15 September 1957, he attended a meeting of several hundred anti-Communists in the Palmgarten Hall of the Frankfurt Botanical Gardens. After a brief speech, he stepped out on to the terrace for some refreshment. Someone handed him a cup of coffee that he had not asked for. He stated that the coffee tasted normal, but for some reason he drank only half the cup (he later put it down to a subconscious action). On leaving the conference, he began to feel desperately tired. 'A strange weight oppressed my stomach and heart' he later wrote. He started to shiver uncontrollably and then fainted. Friends, including Georgi Okolovich, rushed him to hospital, where the initial diagnosis of severe gastritis was made. The doctors in Frankfurt worked around the clock to save Khokhlov's life, pumping him with fluids, vitamins and steroids. The treatment was severe and he was desperately sick. His mouth, salivary glands and throat had dried up to the extent that it was difficult to eat, drink or speak. Of particular concern was the fall in his white blood cell count, which had dropped from the normal range of 4,000–11,000 per microliter of blood to the order of 700. A sample of bone marrow taken from the breastbone confirmed severe suppression of the bone marrow.

The collapse of the blood-forming cells was extremely puzzling since this was not a known effect of thallium poisoning. Blood transfusions were required to restore his blood levels, but the true nature of the poisoning was still elusive.

After subsequent findings of deliberate poisoning had been revealed and released to the press, Khokhlov was transferred to an American hospital for reasons of both security and to receive specialised treatment. His room was guarded by Military police, while six American doctors administered a range of medications, including cortisone, adrenocorticotropic hormone, steroids, vitamins and blood transfusions. The American medical team were not only treating him but making a close study of his condition and his subsequent gradual recovery. By the middle of October Khokhlov had recovered sufficiently to leave the American Hospital and boarded a plane for New York. As he left, the American medical team had still not identified the full list of poisons he had been given, apart from thallium. He had a long and difficult period of recuperation and rehabilitation and whilst convalescing in New York, Khokhlov placed himself in the care of a leading university toxicologist who became interested in his case. His medical history notes were requested, and his full list of symptoms were given further detailed assessment. The symptoms were compared with those exhibited by survivors of atomic bombings, weapons testing and nuclear reactor incidents. The pattern of nausea, vomiting and depletion of blood cells due to destruction of the bone marrow was consistent with recorded observations of radiation poisoning. The combination of thallium, a known metallic poison and radioactivity, the product of the new atomic age came as a complete surprise. The most likely final diagnosis was that he had been poisoned with radioactive thallium in what is considered to be the first case of deliberate poisoning with radioactivity. Thallium-204 is the most likely radionuclide to have been used. This decays with a physical half-life of 3.8 years. In other words, the radioactivity reduces to half its original amount after every 3.8 years. It is produced by placing a sample of stable thallium in the core of a nuclear reactor. This process of producing artificial activity by irradiation in a stream or flux of neutrons is known as neutron activation. The ingenious combination of a toxic metal known to be extremely harmful when ingested, together with the biological effects of radioactivity, provides an extremely potent and deadly poison. It is considered that Russian atomic scientists would have had the ability to produce radioactive thallium at that time, from their relatively new nuclear research facilities. Russia's first nuclear reactor was built in 1954 and the Russian state was keen to demonstrate advances in science and technology. Khokhlov's poisoning had coincided with the successful Soviet launch of the first orbiting space satellite 'Sputnik'. He reflected upon this in his book, when he wrote:

> I, too, was an exhibit of the achievements of Soviet science. Totally bald, so disfigured by scars and spots that those who had known me did not at first recognise me, confined to a rigid diet, I was nevertheless also living proof that Soviet science, the science of killing, is not omnipotent.

Khokhlov continued to work with the US intelligence agencies, and although he was gifted in mathematics, he completed a doctorate in psychology at Duke University and went on to become a professor of psychology at California State University (Figure 1.2). He eventually re-married and had two daughters and a son.

FIGURE 1.2   Nikolai Khokhlov at a US Press Conference after his defection showing pictures of his wife and son.

Source: Nikolai Khokhlov. www.alamy.com/feb-26-2012-nikolai-khokhlov-soviet-spy-murder-squad-image69521903.html

Sadly, his son died of kidney failure when he was quite young. There was serious suspicion in US intelligence circles that his son's illness was the result of a further poisoning incident with an unidentified substance, which was carried out as a revenge for the failed attempt to poison his father. In 1992 he was granted a full pardon by Boris Yeltsin, who was then Russian President. Nikolai Khokhlov died of a heart attack in San Bernardino, California, in September 2007.

Many illness and deaths go unexplained, so it is difficult to know the true number of events that have involved the deliberate exposure of individuals to radiation or radioactivity. The most notorious case of poisoning in recent history is the London polonium poisoning, which took place in 2006 and is described in Chapter 12. However, there are a number of criminal actions that have been reported using radiation sources. Some of these have been listed in Johnson's online database of radiological incidents and related events. There was also a list compiled by Mohtadi and Murshid from the University of Wisconsin, following a review that was supported by the US Department of Homeland Security in 2006. The following record describes some of the documented cases either from scientific publications or from criminal records.

1.8 POISONING WITH TRITIUM (HYDROGEN-3)

In February 1990 routine health monitoring of staff working at the Point Lepreau Nuclear Generating Station in New Brunswick on the east coast of Canada showed elevated levels of radioactivity in urine samples. A disgruntled 33-year-old assistant plant operator Daniel George Maston, who worked at the power plant, took a sample of heavy water from the boron reactor moderator system and poured it into the chilled 5 gallon water dispenser in the staff dining area. Heavy water which is also known as deuterium oxide is not radioactive itself, but it is used to cool the reactor and becomes radioactive as it circulates through the reactor core. Eight employees drank the contaminated water. One individual who was working in a high-temperature area was carrying out stress work, requiring alternating periods of work, rest and rehydration. He consumed significantly more water than the other employees. The incident was discovered when the urine samples from the staff showed elevated urinary levels of tritium, a radioactive form of hydrogen (hydrogen-3). Maston was arrested and during a short court hearing witnesses described him as a 'quiet guy'. A number of workers at the plant thought that this was intended as a bad practical joke rather than a malicious act. The long-term health consequences on the affected staff were considered to be negligible.

A similar incident was discovered in November 2009 at the Kaiga Nuclear Power Generating Plant in Karnataka, on the southwest coast of India. Abnormally high levels of tritium was detected in the urine of 55 plant workers. It is believed that the perpetrator added the tritiated water through the 'overflow tube' of the staff drinking water cooler. Plant managers only admitted that the incident had occurred after it had been reported by the media and management did not report the event to the police for a further week. The Nuclear Power Corporation of India, which operated the Kaiga plant, stated that two workers had received significantly elevated dose. The police commented to the media that they were not convinced that this was the first occurrence of its kind at the Kaiga plant!

Tritium is a radionuclide of hydrogen with a physical half-life of 12.3 years. It decays to helium giving off beta radiation (electrons). It does not pose an external danger to biological tissue as the beta particles cannot travel very far; however, when inhaled or ingested, it can result in internal poisoning. Intake can occur by swallowing, inhalation and via skin puncture or through open wounds. Absorption through the gastrointestinal tract normally takes place within a few minutes and is complete within 45 minutes. Once tritium enters the bloodstream, it quickly disperses and is uniformly distributed throughout the water spaces of soft tissues. It is one of the less radiotoxic radionuclides, and following poisoning incidents it can be readily flushed from the body by increasing water consumption. Tritium is commonly used in civilian and military nuclear power plants and has been observed to be discharged into the environment in relatively large quantities.

1.9 POISONING WITH PHOSPHOROUS-32

Phosphorous is an element that is essential for life. When it combined with oxygen it makes phosphates which hold the cellular DNA (deoxyribonucleic acid) together, make bones strong and is fundamental for the chemical processes within the cells of the body. Phosphorous was discovered by the German chemist Henning Brant around 1670 when he was boiling his own urine to try and make gold! When taken in large amounts phosphorous can kill. Many workers in the early match making factories inhaled large amounts of phosphorous, which resulted in the disintegration of the bones and teeth, a condition called 'phossy jaw'. Radioactive phosphorus-32 is a high-energy beta emitter with a physical half-life of 14.3 days. It is routinely used in biomedical science laboratories for radiolabelling molecular probes and for DNA sequencing. The high-energy beta particles emitted from phosphorous-32 can pose an external risk if the material is splashed on the skin or gets in the eye. Following ingestion or inhalation, phosphorus is taken up into the bloodstream and readily incorporated into bone tissue and can be incorporated into nucleic acids and other essential cellular components. Between 1994 and 1996 a young male graduate student working in the Institute of Plant Pathology at the University of Taiwan was poisoned on about 30 occasions with phosphorus-32 and other materials stolen from the molecular biology laboratories at the university. The radioactive element was placed in the victim's drinking cup and eating utensils by a fellow student. From late 1994 the victim suffered diarrhoea and abdominal pain which was accompanied by poor appetite and weight loss. He later lost most of his moustache. He was informed of the acts by the perpetrator in early 1996 and suffered continuing health effects up to 1999. Estimations of the radiation dose the victim received were made by the assay of radioactivity of his urine. The health effects were considered to be significant and he was kept under continuing health surveillance.

There were two similar incidents in the US in 1995. The first involved Dr Maryann Ma, a post-doctoral researcher working in laboratories at the National Institute of Health. She filed a complaint with the Nuclear Regulatory Commission alleging that she was deliberately poisoned after her supervisor at the National Institute of Health repeatedly urged her to terminate her pregnancy, so she and her husband could complete a cancer research project. The poisoning triggered an enquiry by the Institute's Radiation Protection Office, the Police and investigators from the US Nuclear Regulatory Commission. The report concluded that 26 workers had been affected and that Maryann Ma received the highest amount of radioactivity after someone had placed phosphorous-32 in food and drink stored in the staff fridges and a water cooler in a nearby staff room. The amounts ingested were not considered to pose a significant health risk and no other abnormal levels of exposure were detected. Alarmingly an almost identical case occurred months later affecting Yuqing Li, a student who was found to have consumed around 19 megabecquerels (MBq)[1] of

---

[1] Further explanation on dosimetry, radiation units, such as becquerels and sieverts, and the biological effects of radiation is given in the Glossary and in Chapter 11. The table of unit multiples and submultiples (metric prefixes) is given in Appendix 1.

phophorus-32 at the Massachusetts Institute of Technology Cancer Research Centre in Cambridge, Massachusetts. In both incidents Chinese researchers were poisoned with similar amounts of phosphorous-32. The US Nuclear Regulatory Commission report concluded that these were deliberate acts and that security of materials and radiation protection measures were inadequate.

1.10 IRRADIATION WITH SEALED SOURCES

There are records of a number of attempts by different individuals to cause injury or death by irradiation following the emplacement of sealed radioactive sources. A sealed radioactive source is radioactive material that is permanently sealed in a capsule or bonded in a solid form. Sealed sources are typically small in size ranging from a few millimetres to several centimetres, but they can contain high amounts of radioactive materials such as cesium-137, cobalt-60 or iridium-192 which are mainly used for industrial or medical purposes. For most applications the source is installed in a shielded device that is designed to open a shielded window or to allow the source to move safely out of device where the radiation beam is used and to be safely returned to the device after the operation is complete.

In June 1960 a 19-year-old research worker at a radiological laboratory in Moscow committed suicide by exposing himself to a cesium-137 source. He took a capsule containing the source from the laboratory and put it in his left trouser pocket for 5 hours, then shifted it around his abdomen and back for a further 15 hours. His whole body dose was 15–20 sieverts (Sv). Any dose around 4 Sv or greater is a serious threat to life. He developed symptoms of radiation sickness within hours and died after 15 days.

In April 1972 a man in Harris County, Texas, who had been issued with limited rights of access to his 11-year-old son following the divorce from the child's mother in 1971, obtained two capsule sources of caesium-137 which he was licensed to use by his company for oil and gas prospecting. Each capsule contained 37–74 gigabecquerels (GBq) of radioactivity. From April to October he used the capsules between 5 and 8 occasions to irradiate his son. Around 8 April 1972 during a weekend visit, the father placed the capsules in a pair of headphones and instructed the boy to use them. On April 18 the boy was taken to a doctor for treatment of skin blisters. On two further occasions in July the father put a sedative in the boy's orange juice and put him to sleep on a cushion containing the capsules. On both occasions the boy awoke feeling sick and he developed a rash on his thighs. Again, on a visit in August he was left to sleep on a couch containing the two capsules. By this time he was under the care of a doctor for skin lesions and for hair loss from one side of his head. On a further visit in October the father placed the two capsules on the boy's legs while he was sleeping. After being referred to a plastic surgeon in February 1973 the boy's worsening skin lesions were diagnosed as being radiation induced, prompting further investigation. The boy required 16 operations including skin grafts from January 1974 to November 1978. The irradiations also resulted in functional castration, leaving the boy in need of testosterone replacement. The case was reported to the legal authorities, and on 31 January 1974 after searching the father's house the sources and remote handling tools were found. Charges were brought against the father on May 2 and he was convicted on 17 April 1975. The father appealed but the appeal was denied. After jumping bail, he was finally apprehended in 1981 following an FBI investigation.

In May 1979 an employee at a nuclear reprocessing plant at La Hague, France, attempted to kill his employer by placing a radioactive graphite fuel plug element under the seat of his car. The victim sustained a 0.3 Sv dose to his spinal bone marrow and 4–5 Sv to his testes. The perpetrator was tried and convicted of poisoning by radiation, given a fine of $1,000 and served 9 months in prison. The victim had suffered localised serious exposure but his long-term health outcome was not recorded.

There are at least three known cases of attempted murder using sealed sources that took place in Russia. In August 1991 sealed caesium-137 sources were placed in the office chairs of two company directors at a company in Bratsk, Irkutsk. One of the directors developed symptoms of radiation sickness. In April 1993 radioactive sources thought to be cesium-137 and/or cobalt-60 were placed in the chair of Vladimir Kaplun, director of the Kartontara packing company in Moscow. Over several weeks Kaplun developed symptoms of radiation sickness and was admitted to hospital for a month before he died of radiation poisoning. The source of radiation was only identified by colleagues after his death.

Between February and July 1995 a man living in the Zheleznodorozhny area of Moscow was irradiated by a 48 GBq source of cesium-137 that had been secretly hidden in the door of his vehicle. The driver was exposed over a period of 5 months before the source was discovered. He had sought medical treatment for loss of hair to his thigh and was found to have a low level of red and white blood cells and platelets and loss of sperm. The whole body dose was 8 Sv and his thigh dose was 65 Sv. After 8 months of treatment, his blood cells failed to mature and after 15 months he developed leukaemia. He was treated in hospital periodically from 7 July 1995 and finally died on 27 April 1997.

There is a further recorded case in Guangzhou, China, in May 2002 when the Chinese nuclear scientist Gu Jiming used forged papers to obtain an industrial machine containing iridium-192 pellets. He placed the sources above the ceiling panels in the hospital office of his rival. The victim developed symptoms, including memory loss, fatigue, loss of appetite, headaches, vomiting and bleeding gums. Another 74 staff members of the hospital, including one pregnant woman, also developed symptoms. On being found out, Jiming was convicted in September 2003 and given a suspended death sentence (life imprisonment) and an accomplice was sentenced to a 15-year prison term.

2 ☢

Designer Poisons

Away! Though art poison to my blood.
Cymbeline, William Shakespeare

2.1 POISON LABORATORIES

The motives for poisoning changed from getting rid of unwanted family members, personal gain or malice and became more political when governments started carrying out research into the possibility of using poisons as weapons. Over the past 100 years poisons became more exotic, and the methods of delivery became more sophisticated.

Poisons have been used for the targeted killing or assassination of prominent or outspoken persons, usually for political, religious, military or monetary reasons. The motives for poisoning vary, but they include the human pursuits of power, politics, love, hate and revenge. Assassination is one of the oldest tools of political power. Acts have been perpetrated since ancient times, were recorded in the Bible and have changed the path of history. The shooting of Archduke Franz Ferdinand and his wife in Sarajevo in 1912 is attributed to have been the cause of the First World War. Throughout the Second World War, trained assassins and insurgents carried out targeted killings and assassination attempts. The have been 20 known attempts on US presidents and one British Prime Minister, Spencer Percival was assassinated in 1812.

State-sponsored assassinations have been numerous. Some have been quite open whilst some have been carried out as secret covert operations. Poisoning is a particularly subtle tool. The perpetrator needs to get in close contact with the victim who is usually unaware of the event at the time. Since the act of poisoning or lacing food or drink is surreptitious, this gives ample time for the perpetrator to escape and cross international borders, avoiding both detection and detention. Whatever the motive it is apparent that a great deal of effort and expense has been devoted to the development of the 'perfect poison' by a number of countries including the US, UK, Israel and Russia. Published reports from the US Department of Defence and the Arms Control and Disarmament Agency have suggested that five counties, Iran, Iraq, Libya, North Korea, and Syria, possessed biological warfare programs, although the accuracy of information may be questionable, as evident from the lack of findings in Iraq after the 2003 invasion.

2.2 LABORATORY NO 1

Russia in particular has been accused of some of the most blatant and ingenious poisoning acts. The origin of Russian interest in poisons can be traced back to the start of the Russian revolution. In August 1919 after speaking at a factory in Moscow, Soviet leader Vladimir Lenin was shot twice by Fanya Kaplan, a member of the Social Revolutionary party. Lenin was seriously wounded, but he survived. Kaplan was shot without trial whilst a bloody civil war ensued. In 1920 the czarists were defeated and in 1922 the Union of Soviet Socialist Republics was established. Subsequent examination of the bullets fired during the assassination attempt revealed that they contained curare, a poisonous dark sticky resinous extract from tropical woody plants found in the Americas, notably *chondrodendron* and *strychnos*. In modern medicine curare is known as an alkaloid neuromuscular blocking agent. The South American Indians used this sticky extract to coat the tips of their arrows for hunting. On learning about this it is understood that Lenin was fascinated by the prospect of new killing agents, and under his instruction a poison laboratory known as the 'Special Office' was established in 1921. The special laboratory was to develop products for use against the enemies of the Soviet state who were termed 'the enemies of the people'. In reality this inevitably meant for use against the enemies of the regime. The work of the laboratory was kept top secret. It was only in the 1990s with the demise of the Soviet Union that documentation concerning the activities of the Death Laboratory emerged. Initially the department was under the charge of a number of different experts, who were responsible to the People's Commissariat for Internal Affairs, known as the NKYD. From 1921 the laboratory was led by a professor of medicine Ignatii Kazakov. In 1926 the laboratory was supervised by Genrick Yagoda who became NKYD chief in 1934. During the 1930s, the NKVD was responsible for the political murders of individuals who opposed Stalin. However, in March 1937 Yagoda was arrested on Stalin's orders. He was accused of diamond smuggling, corruption, working as a German agent, attempted assassination of his successor Nikolai Yezhov by sprinkling mercury around his office and of poisoning Maxim Gorky, a soviet writer and his son. Yagoda was shot after his trial. Kazakov, the first head of the 'Special Office', was also executed after being found to be an opponent of Stalin.

Change, fear and mistrust led to a number of accusations and reorganisations. In 1938 Boris Zbarsky, a narcotics expert and professor at the Department of Biochemistry and Analytical Chemistry at the First Moscow Medical Institute, was appointed head of the facility. Several scientists from the Biochemistry Institute collaborated on the research projects. In 1939 the Special Laboratory was split into two special departments for the development of chemical and bacteriological agents. Grigory Moisevich Mairanovsky was appointed head of the chemical facility which was known as Laboratory No 1. Mairanovsky headed several secret laboratories in the Bach Institute of Biochemistry in Moscow between 1928 and 1935. He completed his PhD titled 'Biological activity of the products of interaction of mustard gas with skin tissues' in 1940. This was kept as a 'classified' document. Mustard gas (sulphur mustard) was developed in Europe during the 1800s and became the first chemical weapon of war when it was used by the German Army against British and Canadian soldiers near Ypres in Belgium in 1915. Under Mairanovsky the laboratory became known as the Kamera, which translates in English to cell or chamber, as in *Kamera pytok* or torture chamber. Mairanovsky gained a reputation as a sadistic butcher, equalling the depravity and acts of Joseph Mengelev, the SS doctor at Auschwitz, who took pleasure in selecting subjects for experimentation and supervised the administration of Zyklon B, the cyanide-based pesticide used for the mass killings in the Auschwitz-Birkenau gas chambers.

Mairanovsky and his staff tested candidate substances on prisoners from the Gulag camps and those about to be executed. He took particular interest in bringing people of varied physical condition and age to have a more complete picture about the action of each poison. A number of deadly poisons were tested, including mustard gas, curare, cyanide digitoxin and ricin. In some experiments Mairanovsky observed that a mix of chemicals called 'Injection C' made subjects talk more openly and answer questions, leading to further experiments on the use of hypnotic drugs (Truth Drugs) to be developed for interrogation techniques. The aim of the poisoning experiments was to find a tasteless, odourless chemical that could not be detected at post-mortem. Some poisons caused great suffering, and to mask any shouts from the unfortunate victims a radio was brought in to the Kamera room, to be turned on during the experiments. Candidate poisons were given to the victims as 'medication' with a meal or drink. If the desired effect was not observed Mairanovsky would inject additional amounts. He also personally carried out the execution of some prisoners with a preparation having the desired properties, carbylamine choline chloride which was called C-2 or K-2. According to witness statements after dosing, the victims changed physically, became shorter, quickly weakened and became calm and silent, dying within 15 minutes.

Between 1951 and 1953, a group of predominantly Jewish doctors from Moscow, including Mairanovsk, were accused of a conspiracy to assassinate Soviet leaders. Mairanovsky was arrested in 1951 and colleagues who had worked with him testified describing the nature of his work. He was convicted and served 10 years in prison. It was said that many other lab workers drank excessively, committed suicide or ended up in mental institutions. Laboratory No 1 was officially closed down and then reinstated as Laboratory No 12. Western sources believed these operations expanded to military–industrial proportions to become the largest biological warfare programme in the world, comprising over 40 establishments employing 40,000 workers. Research in poisons, biologicals and delivery devices continued throughout the rest of the 20th century and products and devices were used for targeted clandestine killings both inside and outside Russia. The modus operandi was always to find an opportunity to get close to the victim, administer a virtually untraceable substance and make the victims symptoms appear natural.

It is alleged that Russian scientists continued the development of poisons for military use until the early 1990s, although Russia denies this. One form of chemical nerve agent, in particular, Novichok has been considered to be one of the deadliest ever made. Work on these agents was carried out as part of a Soviet chemical weapons programme codenamed FOLIANT. It is claimed that these agents were designed to be relatively safe to handle but to penetrate NATO standard chemical personal protective equipment whilst being undetectable by conventional chemical detection methods. Novichok, which when translated means 'newcomer' or 'new boy', belongs to a group of organophosphate agents which can cause permanent nerve damage and contraction of skeletal muscles. The first casualty of Novichok poisoning was believed to be Andrei Zheleznyakov, a scientist who came into contact with the residue of an unspecified Novichok agent while working in a laboratory in Moscow in May 1987. After coming into contact with the agent he became unconscious for 10 days. He was treated at a secret clinic in Leningrad for 3 months. The lasting effects of poisoning included chronic weakness in his arms and legs, toxic hepatitis that resulted in cirrhosis of the liver, epilepsy, spells of severe depression and an inability to read or concentrate. His health steadily deteriorated and he eventually died 5 years later in July 1992. The first use of Novichok for deliberate poisoning is considered to have taken place in 1995 when a Russian banker Ivan Kivelid, who was head of the Russian Business Round Table, was murdered together with his secretary, Zara Ismailova. A former business partner of the banker, Vladimir Khutsishvili, was subsequently convicted of the killing, although some Russian historians believe this was one of the first of a series of poisonings against political opponents organised by the Russian security services.

The most high-profile poisoning incident involving Novichok took place in Salisbury in the UK on 4 March 2018. A former Russian military officer and double agent, Sergi Skripal and his daughter Yulia, were found in a state of semi-consciousness on a public bench in the middle of Salisbury. Sergi Skripal was an officer in the Russian Intelligence Directorate (GRU) who was arrested in Moscow in 2004, convicted of high treason and sentenced to 13 years in a penal colony in 2006. He emigrated to the UK in 2010 as part of an Illegals spy exchange deal. The British authorities accused two Russian nationals who travelled to the UK, using the names Alexander Petrov and Ruslan Boshirov, of the poisonings. The two individuals claimed that they came to the UK as innocent tourists to see the spire of Salisbury cathedral; however, they were subsequently identified as being Colonel Anatoliy Chepiga and Dr Alexander Mishkin of the GRU. After full forensic examinations the British authorities concluded that the poison was placed on the outside handle of the Skripal's front door. Twenty-one members of the emergency services and the public were checked for possible symptoms. Two police officers were treated for symptoms of itchy eyes and wheezing and one officer, Detective Sergeant Nick Bailey, who had entered the Skripal's house in a forensic suit was admitted to hospital in a serious condition. He survived but was severely affected. Sergi and Yulia Skripal also survived probably because of degradation of the nerve agent on the door handle due to the damp English winter weather at the time of the act. Over 3 months later on June 30 two people living in Amesbury, 7 miles from Salisbury, were affected by the same nerve agent. Charles Rowley found what he thought was a perfume bottle in a presentation case. He gave this to a close friend Dawn Sturgess, who sprayed it on her wrist. She fell ill within 15 minutes and died in hospital on July 8. Charles Rowley survived after a period of hospitalisation. It was considered that this was an extremely unfortunate event rather than a targeted poisoning. Putting the neurotoxin in a perfume bottle was clearly a way of disguising the contents, enabling the perpetrator to travel across international borders without suspicion and to spray the contents at the intended target site. Discarding the bottle after the deed shows that the perpetrators had a blatant disregard for the health of anyone else who may have found it. Russian use of Novichok as a poison against political opponents surfaced again on 20 August 2020 when Alexei Navalny, a Russian opposition politician, was taken ill after boarding a plane at Tomsk airport in Siberia. The plane Navalny boarded was heading to Moscow but had to make an emergency landing at Omsk where he was taken to hospital. The doctors in Omsk found no evidence of poisoning and the chief toxicologist suggested that his condition might have been caused by stress, excessive dieting or drinking or simply a lack of breakfast! The Russian authorities claimed that Navalny must have poisoned himself or that this was a crime carried out by the Western spies. Navalny was in a coma for 2 days before being flown out to Berlin for specialist treatment. Toxicology tests carried out in Munich found traces of Novichok in Navalny's blood, urine and skin as well as on the bottle he had with him on the flight to Moscow. It is believed that the neurotoxin was placed in his hotel room or in a drink which he had taken at the airport and high levels were also detected in his underwear. From the time of Lenin to present-day Putin, the effects of these blatant, targeted acts of poisoning using exotic agents have baffled doctors and forensic examiners, causing international and political outrage.

2.3 UNIT 731

In 2014 the US National Archives declassified 100,000 pages containing the details of Japanese war crimes. Subsequently in response to a request by Katsuo Nishiyama, a professor at Shiga University of Medical Science, the archives passed on the names of 3,607 people involved with Unit 731, a notorious branch of the imperial Japanese army that conducted experiments on Chinese civilians in the 1930s and 1940s. Unit 731 was originally started to promote public health by conducting research that would benefit Japanese soldiers, such as understanding the ways in which the human body could endure hunger and thirst and fight diseases. The early experiments were carried out on volunteers who had signed consent; however, as war intensified their methods changed and the experiments became more severe. Japan's biological weapons program commenced in the 1930s, partly because Japanese officials were intrigued that germ warfare had been banned by the Geneva Convention of 1925. They reasoned that if it was so dreadful to have been banned under international law, it must make a powerful weapon. Unit 731 and other affiliated units, 1644 and 100, were a part of the Epidemic Prevention and Water Purification Department that was largely coordinated by the Imperial Japanese Army. The work started in the second Sino-Japanese War (1937–1945) to carry out lethal human experiments for research on biological and chemicals warfare. The Japanese Army, which then occupied a large area of China, evicted the residents of eight villages near Harbin, in Manchuria, to site the headquarters of Unit 731. By placing the unit in China, the Japanese had access to research subjects on whom various chemical and biological agents could be tested. The subjects were called *marutas*, or *logs* (taken from felling wooden logs), and most were Asian, Communist sympathisers or criminals. The subjects were chosen to give a wide cross-section of the population. The *Kempeitai*

(Japanese military police) would round up those with suspicion of being common criminals, captured bandits, anti-Japanese partisans, political prisoners, the homeless and mentally handicapped. Throughout the Second World War a strong focus of the work was to examine epidemic-creating biowarfare weapons in widespread assaults against the Chinese military and civilian populations. At least 12 large-scale field trials of biological weapons were performed including 11 Chinese cities. An attack on Changde in 1941 reportedly led to approximately 10,000 biological casualties and ironically the death of 1,700 poorly prepared Japanese troops, mostly from cholera. During 1940 and 1941 plague-infected fleas were bred in the laboratories and spread by low-flying airplanes upon Chinese cities, including Ningbo and Changde in the Huan Province. Military aerial spraying killed tens of thousands of people with bubonic plague epidemics. One expedition to Nanking involved spreading typhoid and paratyphoid into the wells, marshes and dwellings in the city, as well as lacing snacks to be distributed among the population. From the research documents of the epidemics that ensued, the Japanese researchers concluded that paratyphoid fever was 'the most effective' of the pathogens studied.

In the Japanese laboratory experiments, human subjects were used to test injuries received from grenades and flamethrowers positioned at various distances and positions and work was carried out to determine the effects of extreme temperatures, burns, electrocution and exposure to lethal doses of X-rays on human survival. Unlike some of the Nazi doctors who conducted experiments on prisoners, none of the individuals involved with the experiments at Unit 731 were ever punished for their crimes. American forces, chiefly at the behest of General Douglas MacArthur, did not put workers of Unit 731 on trial. MacArthur granted immunity to those involved in exchange for the information they had obtained while doing their experiments. At the end of the war, many of the perpetrators went on to enjoy successful careers in their respective fields.

2.4 EDGEWOOD ARSENAL AND ULTRA

Although the 20th century was a time when a huge effort was made towards human experimentation, sadly, unregulated testing on unsuspecting and vulnerable individuals had taken place throughout American history. These range from the work of J Marion Sims who is often referred to as the 'father of gynaecology' and who performed surgical experiments on enslaved African women without anaesthesia in the 1840s to Dr Leo Stanley, the chief surgeon at the San Quentin Prison. Between 1913 and 1951 Stanley performed a wide variety of experiments on hundreds of prisoners. His specialised interest involved experiments where he removed the testicles from executed prisoners and surgically implanted them into other living prisoners. It has also been revealed that a number of US and Canadian citizens were unwittingly used as test subjects in experiments that were classed as illegal. For most of the Second World War the Russians and Americans fought as allies; however, political differences and state interests led to great distrust. Americans were suspicious of Joseph Stalin's tyrannical rule and the threat of communism. This intensified after 1945 and American officials encouraged the further development of atomic weapons that had ended the final phase of war in the Pacific. A period of intense political and economic rivalry began with numerous threats, spying and espionage. In January 1950, the US discovered that Klaus Fuchs, a refuge from Germany who was a theoretical physicist working for the British mission in the Manhattan Nuclear Bomb Project, had been giving information to the Soviets. At his trial, Fuchs identified a number of individuals who had been passing the information on and this led to the naming of Julius and Ethel Rosenberg, an American couple who had both been members of communist organisations in the US. Julias had worked in the US army Signal Corps, but he was dismissed from the army when they discovered his past membership of the Communist Party. The couple spied for the Soviet Union during the 1940s and provided information on sonar, radar, jet propulsion and the triggering devices for the atomic bomb. They were convicted of espionage and executed in 1953. The receipt of these details enabled the Soviets to build and test their own atomic bomb in 1949. President Truman responded by announcing that the US would build an even more destructive atomic weapon: the hydrogen bomb. The nuclear 'arms race' had begun. A number of regional conflicts known as 'proxy wars' took place around the world in places such as Malaysia, Korea and Vietnam, but the continued development of rocketry and nuclear war heads by Russia and America culminated in the Cuban Missile crisis of October 1962. In Russia the NKVD, which had been responsible for internal security since 1917, transformed into an organisation of secret police known as the Ministry for State Security in 1946. In 1954 the Committee for State Security (KGB, Kremlin Border Guard) was established. The KGB was considered to be the most effective information-gathering network in the world. After the construction of the Berlin Wall East Germany became the focus of ideological conflict between East and West. The East Germans established a domestic secret police organisation, the Stasi, to spy on its own citizens to supress any internal anti-communist movements or actions (See Chapter 12).

The CIA was the main information and spying organisation in the US, and although the Soviets were the main interest, the CIA was also involved in a number of secret operations and regime changes around the world. The British intelligence organisation MI6 often shared intelligence information with the US and other allies. With the ongoing intense ideological and military conflicts it comes as no surprise that the West undertook considerable effort to compete on the scientific and technological application of clandestine subversive methods for the extraction of information and elimination of foreign undesirables and agents. In the 1950s, some officials in the US Department of Defence publicly asserted that many forms of chemical warfare were more 'humane' than existing weapons. For example, certain types of 'psychochemicals' would make it possible to temporarily paralyze large population centres with minimal damage to homes, buildings and infrastructures. According to a testimony by Major General Marshall, the US Army's chief chemical officer, Soviet advances in the chemical and biological weapons were cited as the incentive for giving impetus to research efforts in this area.

After the Second World War the US military obtained information on a number of substances from the Japanese and the formulas of nerve gases developed by the Nazis, notably tabun, soman and sarin. From 1948 the army conducted classified experiments on human subjects at the Edgewood Arsenal in Maryland. These were carried out as a Medical Research Volunteer Programme mainly to assess the effects of psychochemical warfare agents on military personnel as well as the use of pharmaceuticals, vaccines and protective clothing. Over 30 years 7,000 US military personnel and 1,000 civilians were used as subjects to test over 250 chemicals. Over the same period the US Central Office of Intelligence carried out a mind control programme of human experiments called Project Ultra which was to identify and develop drugs and procedures to be used to weaken the individual and force confession through interrogation. The techniques included the covert administration of high doses of psychoactive drugs such as LSD and also involved the use of chemicals, electric shocks, isolation, hypnosis, sleep deprivation, verbal and sexual abuse and torture. The code names for drug-related experiments were 'Project Bluebird' and 'Project Artichoke'.

As with the Russian research, collaboration with medical practitioners and scientists was a common feature of the US experiments. After the Second World War, the US Army worked with Harvard anaesthesiologist Henry Beecher at an interrogation centre at Camp King in Germany. These experiments focussed on the use of psychoactive compounds on human subjects including the debriefing of former Nazi physicians and scientists who had previously worked along similar lines during the war. It is understood that subsequent work involved over 80 institutions in America, including colleges and universities, hospitals, prisons and pharmaceutical companies. This included well respected institutions including The University of Chicago, The University of Michigan, New York University, The University of Pennsylvania, Johns Hopkins University, The Rockefeller Institute of Medical Research and Sloan Kettering Institute.

Experiments of this nature increased dramatically after the Second World War. In 1950 the US Navy sprayed large quantities of the bacteria *Serratia marcescens* (considered harmless at the time, but now known to cause a number of human infections) over San Francisco in a project called *Operation Sea Spray* to simulate a biological warfare attack. Numerous citizens contracted pneumonia-like illnesses and at least one person died as a result. A series of field tests using entomological weapons was carried out during the 1950s by dropping uninfected mosquitoes over the state of Georgia to see if they survived to feed from humans after the drop. These were known as *Operation Big Itch*, *Operation May Day*, *Operation Drop Kick* and *Operation Big Buzz*. Japan had previously used entomological warfare on a large scale during the Second World War in China. Unit 731 of the Japanese Army had previously dispersed plague-infected fleas and flies covered with cholera from low-flying airplanes to infect populations in China. It is estimated that 500,000 Chinese died of disease from the resulting epidemics. In 1966, the US Army released *Bacillus Globigii* into New York Subway tunnels, as part of a field test called 'A Study of the Vulnerability of Subway Passengers in New York City to Covert Attack with Biological Agents'.

Whilst some of these mass poisonings, exposures and infestations were carried out as acts of war, a common feature of the experiments undertaken in individuals is that they were often committed against the vulnerable and the powerless. In an attempt to provide clear rules on what was legal when conducting human experiments, the Nuremberg Code was introduced in August 1947, following the Nuremberg trials of Nazi doctors who were convicted of the crimes of human experimentation on concentration camp prisoners. This stated that it was necessary to obtain the participant's voluntary consent without exception and that they had the right to withdraw from the experiment at any time. This was subsequently superseded by the Declaration of Helsinki which established the modern guidelines for medical research ethics. The statement of ethical research principles was first adopted in 1964 by the World Medical Association to provide framework for physicians and other investigators conducting medical research involving human subjects. Despite signing up to these principles, individuals and countries continued to undertake unethical human experimentation during peace time.

3 ☢

Mysterious Rays

All bodies are transparent to this agent. . .
For brevity's sake I shall use the expression 'rays' and to distinguish them from
other of this name I shall call them X-rays.
Wilhelm Roentgen

3.1 RADIATION, MATTER AND ENERGY

If you ask someone what radiation is, you will probably get a number of different answers. One person will say that it is really dangerous and can give you cancer whilst another person will say it is used to cure cancer. Some will say it is the basis for a nuclear bomb capable of destroying a city and yet others might light heartedly say it can give you superpowers, like the Incredible Hulk (who was exposed to gamma rays) or Spiderman (who was bitten by a radioactive spider). The best way to describe radiation is to say it is energy in the form of waves or particles travelling through matter.

This energy includes electromagnetic radiation, such as radio waves, microwaves, infrared, visible light, ultraviolet, X-rays and gamma rays. Everything we see and feel in the world around us consists of matter. Scientists call this 'normal matter' since they believe that the universe is also made up of two other substances: 'dark matter' and 'dark energy'. At some time in the future this may become as naive as saying that the universe is made up of earth, air, fire and water; however, for the time being it is convenient to run with this scenario. Leaving dark matter and dark energy aside (since this book is not going to try and describe the nature of the universe) normal matter in our world is best described as being made up of atoms. As the saying goes 'never trust an atom as they make up everything'. The early Greek philosophers Leucippus and his pupil Democritus, who lived in the 5th century BC, believed that all matter was made up of various imperishable indivisible elements which he called *atomos* which literally means *uncuttable*. In other words, he believed atoms to be the smallest building block of matter and each element was made up of the same kind of atom. Of this he stated, By convention sweet, by convention bitter, by convention hot, by convention cold, by convention colour: but in reality, atoms and void.

In 1704 Isaac Newton proposed a mechanical universe made up of small solid bodies moving around in motion, but it was John Dalton, the English chemist and meteorologist, who later proposed an atomic theory with spherical solid atoms having measurable properties. As physicists and chemists began to study the properties of matter, electricity and magnetism, different models of atomic structure were developed. The aim was to explain a model for the existence of solids, liquids, gasses, plasma (these being similar to gases but containing highly charged particles) and ultimately space and time. In 1869 the Russian chemist Dmitri Mendeleev proposed that if elements were arranged according to their (atomic) weight, they exhibit an apparent order or periodicity of properties. It is thought that Mendeleev was inspired by a card game similar to Solitaire or Patience. In the game, cards are arranged both by suit, horizontally, and by number, vertically. To put some order into his study of chemical elements, Mendeleev made up a set of cards, one for each of the 63 elements known at the time. In 1873 the Scottish mathematician and physicist James Clerk Maxwell, who was interested in electricity and magnetism, proposed that electric and magnetic fields filled the void between atoms. Subsequent work by a number of notable physicists including Frederick Soddy, Ernest Rutherford, Max Plank, Albert Einstein and Niels Bohr led to a greater understanding of atomic structure and the positioning of each element in Mendeleev's periodic table. The commonly used description of the atom is one of a central nucleus, which contains most of the mass, surrounded by one or more negatively charged orbiting electrons, giving the appearance of a mini solar system with orbiting planets (Figure 3.1). The outer orbiting electrons are only weakly bound to the atom and can be released. In fact, it is the flow of electrons through a suitable material, such as wire, that gives rise to electricity.

The central nucleus contains protons and neutrons (collectively known as nucleons) which are firmly bound together with atomic forces. The protons have a positive charge and neutrons have no charge. Electrons, protons and neutrons are all known as fermions. The electron is by far the smallest of these three particles weighing 9.11×10^{-31} kilograms. Protons have a mass 1,836 times that of the electron, with a weight of 1.6726×10^{-27} kilograms. Neutrons are the heaviest of the three constituent particles, 1,839 times the mass

Atomic model

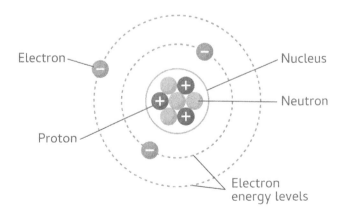

FIGURE 3.1   Schematic model of an atom.

Source: Schematic model of an atom. www.shutterstock.com/image-vector/atomic-structure-model-chart-neurons-protons-1140518345

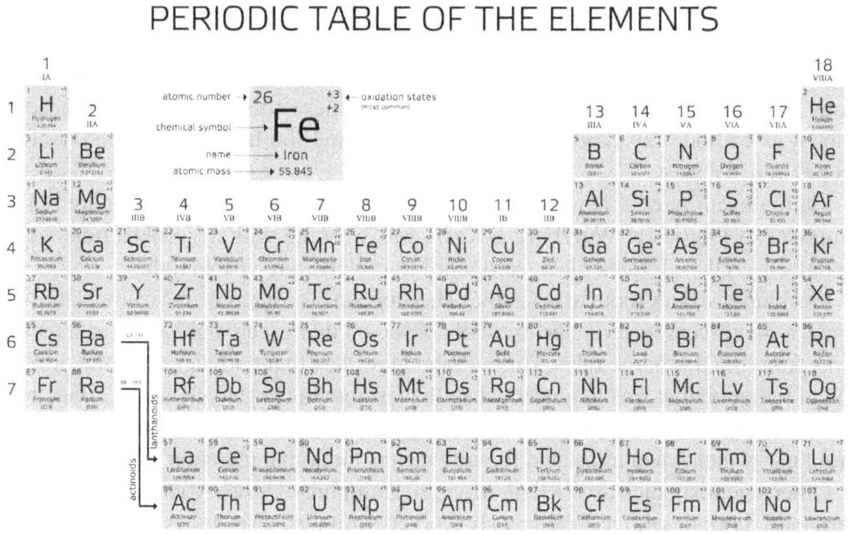

PERIODIC TABLE OF THE ELEMENTS

1	2											13	14	15	16	17	18
H																	He
Li	Be											B	C	N	O	F	Ne
Na	Mg	3	4	5	6	7	8	9	10	11	12	Al	Si	P	S	Cl	Ar
K	Ca	Sc	Ti	V	Cr	Mn	Fe	Co	Ni	Cu	Zn	Ga	Ge	As	Se	Br	Kr
Rb	Sr	Y	Zr	Nb	Mo	Tc	Ru	Rh	Pd	Ag	Cd	In	Sn	Sb	Te	I	Xe
Cs	Ba		Hf	Ta	W	Re	Os	Ir	Pt	Au	Hg	Tl	Pb	Bi	Po	At	Rn
Fr	Ra		Rf	Db	Sg	Bh	Hs	Mt	Ds	Rg	Cn	Nh	Fl	Mc	Lv	Ts	Og

La	Ce	Pr	Nd	Pm	Sm	Eu	Gd	Tb	Dy	Ho	Er	Tm	Yb	Lu
Ac	Th	Pa	U	Np	Pu	Am	Cm	Bk	Cf	Es	Fm	Md	No	Lr

(atomic number → 26, oxidation states +3 +2, chemical symbol → Fe, name → Iron, atomic mass → 55.845)

FIGURE 3.2 Periodic table of the elements.

Source: Periodic table of the elements. www.shutterstock.com/image-vector/periodic-table-elements-atomic-number-weight-529925884

of the electron, at 1.6749×10^{-27} kilograms. In the last 50 years particle physicists have discovered a number of smaller elementary particles such as quarks, leptons and bosons. Although these have helped physicists to understand what makes up the universe, it is possible to give a perfectly adequate explanation of the nature of matter, radiation and radioactivity without reference to the elementary particles. So, all descriptions here shall be limited to atoms, electrons, protons and neutrons. The sum of protons and neutrons is known as the mass number and given the symbol 'n'. The number of protons in the nucleus is the *atomic number* of the element (Z) and this is a unique number for each element. The periodic table of the elements shows the position of each element listed in order of increasing atomic number Z. As of 2016, the periodic table shows 118 known elements, from hydrogen which is a gas with an atomic number of 1 to a super heavy radioactive metal called oganesson (of which only a few atoms have been made) with an atomic number of 118. The number of protons in an atom is normally equal to the number of electrons, thus balancing the atom's overall positive/negative charge, but atoms can have different numbers of neutrons and hence a range of mass numbers. These are known as different isotopes of the element. It is worth stressing that isotopes are not all radioactive. Those that are are referred to as radioisotopes or radionuclides.

3.2 THE WONDROUS ROENTGEN RAYS

The discovery of radiation probably began with the work of William Crookes, the very same person who discovered the element thallium. Crookes had developed a glass tube (Crookes tube) which had air withdrawn to create a partial vacuum. A voltage applied across two electrodes placed at each end of the tube, one positive (anode) and one negative (cathode), causes a stream of negatively charged electrons also known as cathode rays to pass along the length of the tube from the cathode to the anode. During the 1890s these tubes were being used for a number of physics experiments at laboratories around the world. In 1879 Crookes found that the electrons could be deflected using magnetic fields. These tubes subsequently became known as cathode ray tubes and eventually became the basis for the first oscilloscopes and TV sets.

On the evening of 8 November 1895 Wilhelm Conrad Roentgen (Figure 3.3), a Professor of Physics at Wurzburg Bavaria was working with a Crookes tube which he had covered with thick black cardboard to exclude any visible light.

Working in a dark room using a thick sheet of paper covered with barium platinocyanide, a fluorescent material, he excitedly observed fluorescent light as far as 2 metres away from the discharge tube. This strange glow was found to be due to the production of radiation that was produced when the fast moving negatively charged electrons passing along the tube hit the positively charged anode. This effect was termed 'bremsstrahlung' which when translated literally means 'breaking radiation' (Figure 3.4).

Other scientists such as William Morgan, Humphrey Davey, Michael Faraday and Fernando Sanford who had previously worked with discharge tubes had unwittingly

FIGURE 3.3   Wilhelm Roentgen.

Source: Wilhelm Roentgen. www.shutterstock.com/image-photo/wilhelm-roentgen-18451923-german-physicist-received-242820439

X-ray tube device

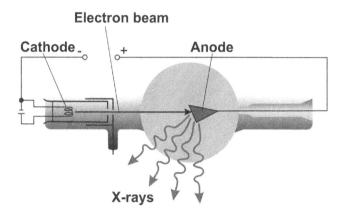

FIGURE 3.4 Diagram of an X-ray tube.

Source: Diagram of an X-ray tube. www.shutterstock.com/image-illustration/xray-tube-device-1373889890

observed the glow from X-rays, but it was Roentgen who first considered these emissions as rays because they were seen to travel in straight lines, producing shadows when passing through more dense materials, such as metal and even bone. He finally called them X-rays, because of their unknown nature. At the time of Roentgen's discovery, Crookes discharge tubes were standard apparatus that had been in use in many laboratories around the world for around 10 years. So, when Roentgen announced that these tubes produced mysterious rays that could pass through solid objects such as books and the human body it was easy for other scientists to replicate the findings. Roentgen discovered the potential medical use of X-rays when he took a picture of his wife's hand and recorded it on a photographic plate. The photograph of Bertha's hand shown in Figure 3.5 was the first radiograph of a human body part.

Roentgen was eager to find out what his wife thought of this new application. When she saw the post-mortem-like image of the bones of her hand with the shadow of the metal ring on her finger, she said 'I have seen my death'. The record of this comment is of interest, since in the heady days of those momentous discoveries the wives of physicists seem to have been largely ignored. According to Bertha, her husband slept and ate in his laboratory, was morose and abstracted and 'resented the intrusions of mundane matters'.

Soon after Christmas in 1895 Roentgen submitted a preliminary communication 'Ueber Eine Neue Art von Strahlen' (A new type of rays) to the Transactions of the Wurzburg Physical-Medical Society and the article was printed immediately. The subsequent spread of the news of these new 'X-rays' was phenomenal. It has been said that it is difficult to think of any discovery in medicine that has spread throughout the world with such speed, apart from the news of the first heart transplant carried out in South Africa by Christiaan Barnard in 1967. The rapid adoption of medical X-ray imaging that followed was all the more remarkable, since the use of effective communications media was not widespread until nearly a century later. The year after Roentgen's

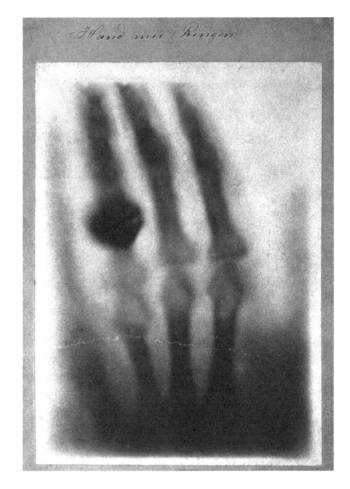

FIGURE 3.5   The radiograph of Roentgen's wife's hand taken in 1895.

Source: X-ray graph of Roentgen's wife's hand 1895. www.ncbi.nlm.nih.gov/books/NBK476355/figure/ch4.f4/ Courtesy of the Wellcome Trust Library, London CC BY-NC 4.0. Image freely available under a Creative Commons (CC-BY-NC-ND) license

announcement, over 1,000 articles on X-rays had been published. The first British report of Roentgen's discovery appeared in the *Daily Chronicle* on 6 January 1896. This was a rather sketchy article which misspelt his name as 'Rontgen' giving an inaccurate description of his apparatus but predicting that this may be an 'excellent expedient for surgeons'. The first scientific report was a short note in *The Electrician* on 10 January 1896. The writer, who thought this would be easily explained, wrote 'that the whole phenomenon seems likely to admit of ready explanation' and ended with a prediction that 'there are few persons who would care to sit for a portrait for 2 hours which would only show the bones and rings of the fingers'. This comment was typical of many others at the time when the practical applications of X-rays were thought to be mainly in photography. Not long after this, advertisers were promoting 'X-ray proof underclothing – especially made for the sensitive woman' (Electrical World, 1896). In 1901 Wilhelm Roentgen received the first Nobel Prize in Physics for his work.

Crookes tubes were of a relatively simple design, essentially not being much more than a super voltage light bulb, and since they were widely available throughout the world, they were easily installed in laboratories and hospitals. The first medical use of X-rays was carried out by John Hall-Edwards in Birmingham, England, on 11 January 1896, when he radiographed a sterile needle which had become embedded in the hand of a colleague. A month later he took the first radiograph to direct a surgical operation and became the first to take an X-ray of the human spine. One of the first radiology departments in · the world was set up at the Glasgow Royal Infirmary in 1896. The head of the department Dr John Macintyre produced a number of remarkable X-ray images: the first X-ray of a kidney stone, an X-ray showing a penny in the throat of a child and a series of images showing a frog's legs in motion. In 1899 Hall-Edwards was made the first Surgeon Radiographer at the Birmingham General Hospital and in February 1900 he joined the Warwickshire Regiment to work as the first military radiographer, during the Boer War in South Africa.

For over 100 years the science and technology of X-ray imaging has revolutionised the way modern medicine is practiced. Physicians and surgeons have had the ability to see inside the body, diagnosing disease, planning treatments and monitoring follow up without resorting to cutting the patient open. As the understanding of the interaction of X-rays with tissues increased the medical profession could balance the relative risk of radiation exposure and treatment. By the summer of 1896 X-ray apparatus was installed in several London Hospitals. Robert Jones, a pioneering Welsh orthopaedic surgeon, reported on the use of the new X-rays (using an exposure of 2 hours) to locate a bullet in the wrist of a 12-year-old boy. The absorbed radiation dose of the X-rays used in this case has been estimated to be 12,500–25,000 times greater than the amount used today. Given the high exposures and lack of shielding it is not surprising that adverse effects of X-ray radiation began to be observed, although it took some time to be appreciated and this was not without misadventure, error and tragic mistakes along the way.

In the US in late summer of 1896 Mr Herbert Hawks (Assistant to Dr Michael Pupion of Columbia) was demonstrating the use of X-rays in Bloomingdales Store, New York, presumably as a sales event to attract customers. After repeated exposures on himself he described severe 'burning' – 'like bad sunburn' on the skin of his hands – causing him to stop work for 3 weeks. Also, in the US in 1896 a particularly unfortunate case was described by Professor Jones of the University of Minnesota Physical Laboratory. He was approached by William Levy who had been shot in the head by a bank robber 10 years previously. The bullet entered his skull just above the left ear and Levy wanted this localised and removed. Prof Jones was reluctant to undertake the investigation but was persuaded to carry out the procedure. On July 8 Levy sat through a 12-hour X-ray exposure from 8 am until 10 pm, with the tube over his forehead, in front of his open mouth. The next day his entire head was blistered and over the following days became red and sore. His lips became swollen, cracked and bleeding, his right ear doubled in size and he lost the hair on the right side of his head. Apart from that, the radiograph successfully showed that the bullet was about an inch beneath surface of the skull and not likely to cause him any harm.

3.3 ILL HEALTH AND SAFETY

Electricians and physicists speculated on the nature of these new and strange rays, some thinking them to be vortices in the ether and Thomas Edison suggested them to be acoustical or gravitational waves. We now understand them to be electromagnetic waves, also known as *photons* or *quanta*. Light, radio waves, X-rays and gamma rays are all electromagnetic radiation. These all possess characteristics of wavelength and frequency but perhaps the most important property of electromagnetic waves from a biological point of view is that they contain energy, which is measured in electron volts (eV). The energy of X-rays used for medical diagnosis today is generally in the range between 40 thousand and 140 thousand electron volts (40–140 kiloelectron volts, keV). Light waves are visible to the human eye, but as the frequency of the waves increases above the visible range there is a corresponding reduction in wavelength, through ultraviolet, X- and gamma rays. With increasing frequency there is also an increase in energy which allows the waves to penetrate through solid materials. Depending on the amount of energy in the wave and the density of the matter, some waves can pass through unaffected whilst some may be blocked. This is the basis of X-ray imaging, where the final image created on an X-ray plate or a detector is formed by the differential absorption and transmission of the rays, resulting in a shadow graph picture. When X-rays pass through the human body, the ones that are stopped and absorbed deposit their energy in the tissues, and this can produce harmful biological effects. Absorbed radiation dose is measured in terms of the energy deposited per unit mass of tissue. Since energy is measure in joules and mass is measured in grams, radiation dose is expressed in terms of joules per kilogram (J kg^{-1}), where 1 J kg^{-1} is called a gray (Gy), named after the English radiation physicist Louis Harold Gray who worked at Mount Vernon Hospital in the UK from the 1930s to the 1950s.

The Gray (Gy) and the Sievert (Sv)

Two units are used to express the amount of radiation absorbed by the body.

The gray is the amount of energy absorbed by an organ or by the whole body. The dose of one Gy corresponds to one joule of energy delivered to one kilogram of tissue.

The sievert is the unit of the effective dose and is used to express how much risk the radiation presents to a person in terms of inducing cancer or genetic damage.

Examples

0.005 mSv, dental X-ray

0.1 mSv, chest X-ray

1 mSv, 1 in 20,000 lifetime risk of fatal cancer

3 mSv, average annual dose to individuals from background radiation

5 mSv, X-ray CT scan of the abdomen

10 mSv, 1 in 2,000 lifetime risk of fatal cancer

1 Sv or greater in a short time period, immediate harm to health

4 Sv, 50% chance of survival with appropriate medical care

To convert a dose from gray to sievert it is necessary to account for the different types of radiation (e.g. X-ray, gamma ray, beta particle, alpha particle and neutron) and the sensitivity of the individual organs to radiation. These are given weighting factors to account for the differences in radiation severity and the differences in the radiation sensitivity of the tissue.

Sievert value = (gray value) × (radiation weighting factor) × (tissue weighting factor)

Further explanation is given in Chapter 11.

The introduction of X-ray imaging quickly revolutionised medical diagnostics and provided new forms of treatment. They were the new and sophisticated way of probing the very substance of living bodies and tissue. By 1905 most hospitals had an X-ray room or X-ray lab, although these tended to be in dark rooms located in basements. The early facilities were often hazardous with high-voltage cables and bare wiring presenting a real risk of electrocution. The X-ray tubes were poorly shielded and would often become very hot. At the time the X-ray pioneers had little or no understanding of any potential dangers of radiation. They did not see a need to protect their patients or themselves from exposure. With fluctuating power sources and poorly focused beams, doctors, equipment operators, patients, bystanders and even those people in adjacent rooms were irradiated. During the early years of use there were many victims of the unseen dangers. Machine operators often tested that their equipment was working by placing their hands in the beam and the damaging effects soon became apparent (Figure 3.6).

It is thought that one of the first fatalities due to X-ray exposure was Clarence Daly, the chief assistant to Thomas Edison, who had many 'radiation burns' on his face, hands and fingers. By 1902 he had developed skin cancer and had to have both arms amputated before dying 2 years later. Edison soon developed a distrust of these new rays.

Mihran Kassabian was an early American radiologist who was born in West Asia and after studying medicine and theology in London he moved to the US taking up a position as director of the Roentgen Ray Laboratory at Philadelphia General Hospital in 1902. His first published paper described the irritant effects of X-rays on the skin. Serious radiation burns to his hand in 1902 led to the development of tumours which resulted in need for amputation of his fingers 6 years later. Dr Kassabian kept a journal with photographic records of his condition, as shown in Figure 3.7. He died of metastatic disease in 1910.

It is widely considered that the first woman to die from working with X-rays was Elizabeth Fleischman. By 1897 after completing a course in electrical science she borrowed enough money from her father to purchase X-ray apparatus and a fluoroscope and set up an X-ray laboratory on Sutter Street in San Francisco. She soon developed skills in both anatomy and photography and performed examinations on patients referred by local physicians. To put her clients at ease she would often expose herself to X-rays to show them that the procedure was painless. She also examined animals and various inanimate objects. In 1898 she provided radiography services to the US Army, where she took what was to be her most famous radiograph,

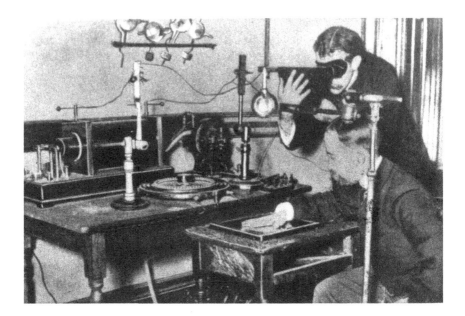

FIGURE 3.6   An early method of testing the output of X-ray tubes using a fluoroscope. The hand could be viewed through a fluoroscope without developing time-consuming X-ray photographs.

showing a Mauser 7 millimetre bullet lodged in the brain of John Gretzer Jr of the 1st Nebraska Volunteers, who was wounded fighting in the Spanish American war. By 1903, the cumulative effects of 7 years of working 12-hour days with unshielded X-ray equipment began to have its effects. She first developed radiation dermatitis on her hands which she attributed to the use of chemicals for developing photographic plates. A year later the skin of her hands became hard and dry and cracked easily due to the secreting glands and hair follicles having been destroyed. The fingers of both hands were found to be badly ulcerated with neocarcinogenic warts. Despite this she continued to work and her doctors made a failed attempt to remove a tumour on her right hand. By January 1905 her entire right arm, scapula and clavicle had to be amputated. In 1905, aged 38 years, she died of lung metastases. Such was the fate of many early radiologists. In 1946 a statistical study of obituaries in the *New England Journal of Medicine* by Dr Helmuth Ulrich found the leukaemia rate among radiologists to be eight times that of other doctors. Other causes of death from multiple myeloma, aplastic anaemia, stroke and heart disease was significantly higher than those in other doctors. It is not surprising that some radiologists were making the call that the time has come when the abuse of the God-given energy should be controlled.

Reddening of skin or skin erythema was the most common early radiation effect observed in both patients and radiologists. In the absence of any reliable objective physical measure, the *erythema dose* was the most popular way of measuring the

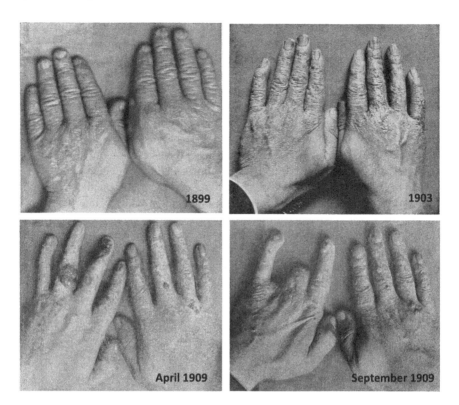

FIGURE 3.7 Photographs of the hands of Mihran Kassabian, a radiologist in Philadelphia who kept a photographic record of radiation damage following repeated doses from calibrating the X-ray tubes. 1899: before any inflammation occurred. 1903: showing chronic X-ray dermatitis. April 1909: showing multiple ulcerations. September 1909: following the amputation of two fingers

Source: Photographs of the hands of Mihran Kassabian. Open source archive,courtesy of the Frances A Conway Library of Medicine and the OpenKnowledge Commons and Harvard Medical School. https://archive.org/details/rntgenrayselectr00kass/page/n9/mode/2up

amounts of radiation exposure. Hospital X-ray machines were calibrated according to the irradiation time required to induce erythema. It was common practice to expose a human thigh at a given distance, voltage and current until the skin began to redden. The *erythema dose* was adopted as the first recognised permissible dose or dose limit.

It is of interest that within a year of Roentgen's discovery an American engineer Wolfram Fuchs, who was working with X-ray tubes, gave what is generally recognised as the first safety advice. This was noted by Clarke and Valentin who wrote the history of the International Commission on Radiological Protection in 2009. This early advice was:

- Make the exposure as short as possible;
- Do not stand within 12 inches (30 centimetres) of the X-ray tube
- Coat the skin with Vaseline (petroleum jelly)

Few workers took any notice of this at the time; however, this advice turned out to be the basis of the fundamental principles of radiation protection that form part of the basic safety measures in use today. These are time, distance and shielding. Minimising the time of exposure, increasing the distance from the source and introducing a physical barrier will all contribute to dose reduction. Vaseline was considered to be helpful in minimising the skin effects, although provided little benefit apart from soothing the skin. The mechanisms concerning the interaction of radiation on tissue were not understood but once the effects had been observed, physicians and radiologists began to take measures to protect themselves, mainly by the use of shielding such as wearing lead-lined vests, helmets and gloves (Figure 3.8).

Radiology grew from a few hundred enthusiasts to an elite medical speciality of thousands of doctors, trained in the use of radiation and the interpretation of radiographic images. To support this growing medical speciality, it was necessary to employ radiographers who were trained in X-ray technology and anatomical positioning.

The first international recommendations on radiological protection were published in 1928 by the newly formed International X-Ray and Radium Protection Committee, which was later to become the International Commission on Radiological

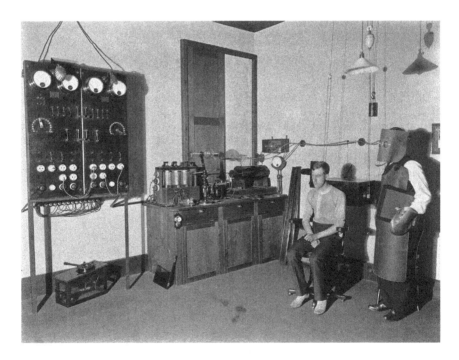

FIGURE 3.8   X-ray room at Johannesburg Hospital, South Africa, circa 1905, showing a radiologist wearing a lead apron, face mask and gloves whilst holding an X-ray plate.

Protection. The scientific fascination with radiation and radioactivity led many radiation biologists to investigate the effects of X-ray exposures on animals and plants. Most early experiments were of poor design and of a qualitative nature, but in the latter part of the 20th century advances in instrumentation for radiation detection and monitoring made quantitative measures possible. As the fields of radiation biology and medical physics became established the understanding of the biological effects of radiation increased. The epidemiological study of exposed individuals and populations has provided data for dose risk stratification, allowing the assessment of the risks and benefits of medical exposures. Today medical X-rays are essential for medical diagnosis and treatment and are a significant source of man-made radiation exposure.

4

Radiant Health

*Passage of the rays for an hour through the head of a laboratory boy of medium
intelligence did not in my hands cause deterioration or improvement thereof.*
Edward Waymouth Reid, Professor of Physiology,
University College, Dundee 1897

4.1 X-RAY VISION

Whilst the medical professions were discovering the beneficial and harmful attri-
butes of Roentgen's discovery, the rest of the world embraced them as a rather curi-
ous and intellectual source of entertainment and amazement. Occultists, spiritualists
and supernaturalists were quick to embrace Roentgen rays as the definitive proof of
the powers they had claimed for decades, even describing them as Thought Rays.
Henri Antoine Jules Bois, the French writer and occultist and founder of the secret
society The Hermetic Order of the Golden Dawn (make what you wish of that!), suggested in the French occult journal *Revue Spirite* that the discovery of the famous
X-rays that traverse opaque materials may well put us on the road to a rational expli-
cation of clairvoyance. The notion of X-ray vision first considered by Bois is still
widely expressed and is a well-known power of the comic superhero, Superman.
Despite the early reports of the adverse effects of exposure to X-rays, people took
their friends and families out to studios to have bone portraits taken. People would
wait in long queues at exhibitions, circuses and shows to see X-ray and fluoroscopy
demonstrations for entertainment and amusement (Figure 4.1). In May 1896, in New
England, Professor Frank Austin used a portable X-ray machine to photograph children's hands as a treat for his daughter's birthday party.

   To some people the possibility of seeing under 'cloak and gown' was seen as a
convivial and saucy encounter. However, to others the penetrating nature of X-rays
allowing people to see beneath clothing was considered to be personally invasive, as
noted in the rhyme below.

X-ACTLY SO!

The Roentgen Rays, the Roentgen Rays,
What is this craze?
The town's ablaze
With the new phase

Of X-ray's ways.
I'm full of daze,
Shock and amaze;
For nowadays
I hear they'll gaze
Thro' cloak and gown – and even stays,
These naughty, naughty Roentgen Rays.

Wilhelma, Electrical Review, 17 April 1896

FIGURE 4.1 Early 20th-century exhibition poster advertising X-ray photographs for public entertainment.

Source: X-ray poster. Image in the public domain

4.2 ELECTROMECHANICAL MEDICAL APPLIANCES

To put all this into context, it should be appreciated that early radiology was being applied in the late 19th and early 20th centuries, which were times of great invention and innovation: a time when the combination of pseudoscience and ignorance led to some bizarre practices and quack medical devices suitable for gullible individuals who were seeking a magical cure for all illnesses. A range of mechanical and electrical devices that did not emit X-rays or radioactivity were promoted and used by medical quacks and charlatans. These could be purchased for the treatment and 'cure' of a wide range of ailments. This was the era when phrenology was used to assess character and electropathy and galvanic power belts were advocated for treating rupture, back pain, kidney and bladder disorders. Dr McLaughlin's electric belt could cure pain, weakness and nervo-vital derangements, and The Boston Medical Institute in Chicago sold an electric belt that they claimed could even improve sexual function! (Figure 4.2).

The Medical Battery Company Limited in London sold 'his' and 'hers' belts that were scientifically constructed to restore new life and vigour to weak men and delicate ladies (Figure 4.3).

The Pall Mall Electrical Association sold a range of Dr Scott's astonishing electrical cures including an electric flesh brush to cure headache, to cure neuralgia and to prevent dandruff and baldness (Figure 4.4). Dr Scott also produced a patented unbreakable electric corset which was designed on scientific principles, generating an exhilarating current to treat the whole system. The inventor claimed that this was 'approved by Parisian models, so adding a graceful and attractive figure to the wearer, whilst applying electromagnetism, the therapeutic value is unquestioned and quickly cures in a marvellous manner, nervous debility, spinal complaints, rheumatism, paralysis, numbness, dyspepsia, liver and kidney troubles, impaired circulation, constipation and all other diseases particular to women and those of a sedentary nature'. Each corset was packed in a 'handsome box' accompanied by a silver-plated compass so that the magnetic influence of the corset could be tested.

Dr Elisha Perkins (no relation to the author) patented a pair of metallic tractors which were described as 'Connecticut's cure for everything that ails you'. These were just a pair of metal rods, one steel and one brass, that 'when passed over the surface of the skin had the power to draw out noxious electrical fluids that lay at the root of suffering'. The reputed benefits of these rods were so convincing that it is understood that even George Washington bought a pair! Glowing testimonials of the Tractors spread to Europe. Perkins' son Benjamin travelled to England where he published an article on 'The Influence of Metallic Tractors on the Human Body in Removing Various Painful Inflammatory Diseases, Such as Rheumatism, Pleurisy, and Some Gouty Affections'. In 1809, the English poet Lord Byron published a satirical rhyme aimed at modern society where he held up the Tractors to ridicule:

Thus saith the Preacher: 'Nought beneath the sun
Is new,' yet still from change to change we run.
What varied wonders tempt us as they pass!
The Cow-pox, Tractors, Galvanism, and Gas,
In turns appear, to make the vulgar stare,
Till the swoln bubble bursts – and all is air!

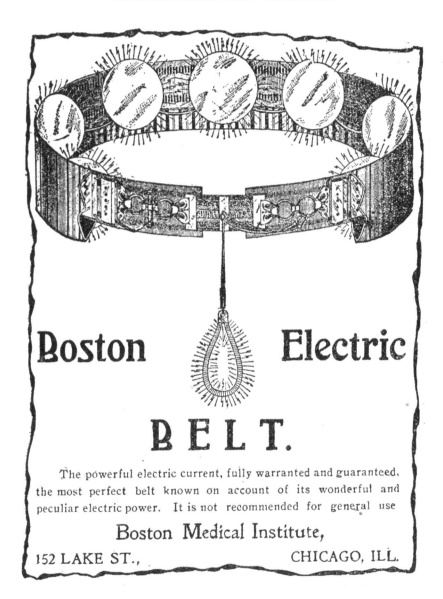

FIGURE 4.2   Advertisement for the Boston Medical electromechanical belt with suspensory for improving sexual function, circa 1890.

Source: Boston electric belt. Courtesy of the Wellcome Trust Library collection, London CC BY-NC 4.0. https://wellcomecollection.org/works/q2z8gnyf

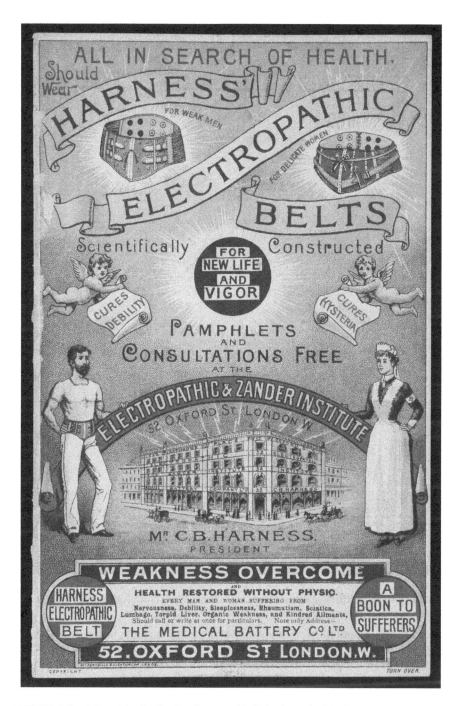

FIGURE 4.3   Advertising leaflet for electropathic belts from the Medical Battery Company, London, circa 1890.

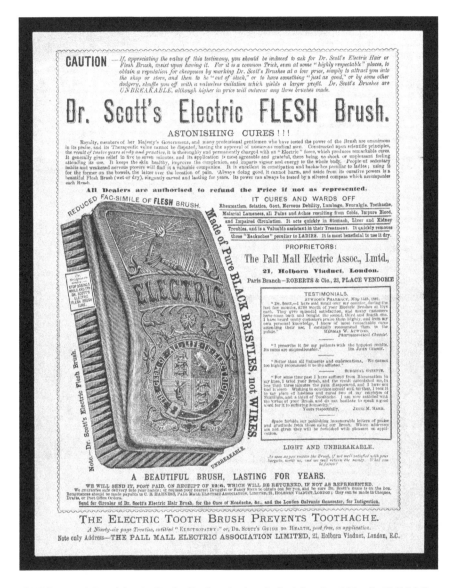

FIGURE 4.4 Advertising leaflet for Dr Scott's electric flesh brush sold by the Pall Mall Electric Association Ltd, circa 1881.

Source: Advertising leaflet for Dr Scott's electric flesh brush. Courtesy of the Wellcome Trust Library collection, London CC BY-NC 4.0. https://wellcomecollection.org/images?query=electrical+devices#

A range of other quirky devices were also promoted, including the self-operating mechanical chiropractor which resembled a modern version of the medieval rack and a number of rectal devices such as the *Prostate Warmer* and the *Recto Rotor* which was intended for private use at home. The manufacturer claimed that when inserted into the rectum this appliance hit the vital spot, for the quick relief of piles, constipation and prostate trouble!

4.3 IF THE SHOE FITS

Since X-rays can clearly show the bones in the body, it is not surprising that someone should think of using them to see how well feet fit into shoes. Dr Jacob Lowe, a Boston physician, is considered to be the inventor of first fluoroscopic device for X-raying feet. During World War I there were many injured military personnel with feet problems and this device reduced the time of consultations by eliminating the need for his patients to remove their boots. He named the device the *Foot-O-Scope* and the company that he formed to produce it was the X-Ray Foot-O-Scope Corporation. Although he filed his patent in 1919 this was not granted until 1927. Two other claims to be the first to develop such a device were made: one by Matthew Adrian who worked for an X-ray firm known as the Milwaukee X-Ray Laboratory and had a patent granted in 1927 and the other by Clarence Karrer who in 1924 was working with his father, a dealer in surgical supplies and X-ray equipment in Milwaukee. At the time it is said that a number of radiologists backed by the Radiological Society of North America asked Karrer to stop using this apparatus since it 'lowered the dignity of the profession of radiology'. In 1926 a patent was granted in the UK for a similar device known as the Pedoscope and was made by the Pedoscope Company Ltd in St Albans. The device was essentially a lead-lined oak cabinet about 1.2 metres (4 feet) tall with one or more viewing windows, resembling something between a Nickelodeon and a 'What-the-Butler-Saw' machine. The customer would stand on a recessed platform above an X-ray tube that was pointing directly up though the feet to a screen at the top. By leaning forward and looking down into a window in the top of the cabinet an X-ray image of the shoe surrounding the bones of the feet could be seen on a fluorescent screen. The radiation level could be changed by selecting one of three settings using buttons marked 'man', 'woman' and 'child' See Figures 4.5 and 4.6.

This was claimed to be a scientific shoe fitting at its best and considered essential for sizing children's shoes allowing the feet plenty of room for growth. In practice it was considered to be a sales gimmick, but although the fluoroscopes were relatively expensive it was considered worth the cost, since customers went out of their way to find a shoe shop where they could view the best fitting shoes. Although the X-rays were directed at the feet the machines produced a large amount of scattered radiation which exposed the pelvic organs and gonads as well as anyone standing nearby. Children and the shop sales staff are considered to be those who were particularly at risk. Radiation dose reconstructions have been made to estimate individual radiation exposures from the shoe fluoroscopes but since there were never any records of the individuals affected it has been difficult to carry out any epidemiological follow up of delayed effects. There are a few case reports, including the case of a shoe saleswoman who developed chronic radiation dermatitis on her own feet and one report of a 72-year-old lady with a basal cell carcinoma of the foot attributed to regular use of shoe-fitting fluoroscopy. In 1957, Pennsylvania became the first US state to ban the shoe-fitting fluoroscope. The following year, the Home Office in the UK required all Pedoscopes to display a safety notice stating that 'Repeated exposure to X-rays may be harmful. It is unwise for customers

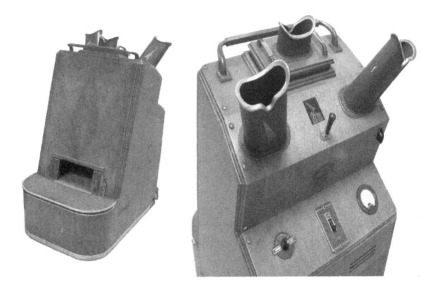

FIGURE 4.5   The Adrian X-ray shoe-fitting fluoroscope. This unit had a 50 kv X-ray tube operating at 3–8 milliamps and was used in shoe stores in the 1930s, 1940s and 1950s.

Source: The Adrian X-ray shoe-fitting fluoroscope. This unit had a 50 kv X-ray tube operating at 3–8 milliamps and was used in shoe stores in the 1930s, 1940s and 1950s. Printed with permission from the Oak Ridge Associated Universities Health Physics Historical Instrumentation Collection www.orau.org/ptp/museumdirectory.htm

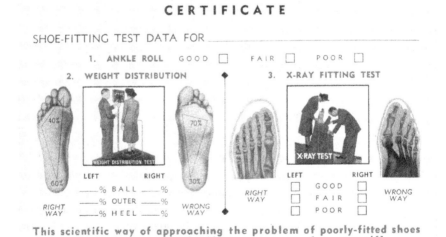

FIGURE 4.6   Shoe-fitting data card for recording the results from the X-ray fitting test.

Source: Shoe-fitting data card for recording the results from the X-ray fitting test. Printed with permission from the Oak Ridge Associated Universities Health Physics Historical Instrumentation Collection www.orau.org/ptp/museumdirectory.htm

to have more than twelve shoe-fitting exposures a year'. However, since records of who had their feet measured in this way were never kept, the limits were never enforced. During the 1970s these machines were phased out in the UK and other countries.

4.4 BEAUTY AND THE BEAST

As with the electromechanical devices, the early use of X-rays in medical practice was unregulated. As the number of X-ray procedures increased, so did the number of people operating the equipment, most of whom had little knowledge of what they were doing. Many of the early operators of X-ray machines were those who had been promoting the use of electromechanical devices. With advances in the technology and equipment design the X-ray generators became more powerful offering not only better quality images but also having the potential to deliver more radiation into the body. In 1896 the physician Wilhelm Marcus, from Berlin, reported dermatological changes in a 17-year-old male who had taken part in various radioscopic procedures. As a result of his report, physicians began to use X-rays for the therapeutic treatment of hypertrichosis (abnormal growth of hair), eczema (dry and cracked skin) and mycoses (fungal infections). Doctors at the time thought that removing the hair with X-rays was preferable to other, more painful techniques such as forceps or hot wax. In 1926 an article appeared in the medical journal *Lancet*, entitled 'X Rays as Depilatory', advocating the use of X-rays for the removal of beard hair rather than wet shaving. In the US adverts appeared claiming X-rays as being the modern scientific method to remove superfluous hair, without the use of electric needles of chemicals. At a meeting of the New York Physical Therapy Society in 1908 Albert Geyser had previously demonstrated a new Cornell X-ray tube named after the Cornell University Medical College, where he was based. The heavy lead glass tube had a small flint glass window to filtrate the beam and reduce the area of exposure to the size of the lesion to be treated. Geyser claimed that the Cornell tube eliminated many of the dangers associated with the use of X-rays including X-ray burns, prompting *The New York Times* to write a story entitled 'New Tube Robs X-ray of Danger', which was clearly not the case. Geyser's experimentation with X-rays on his own hands resulted in lesions which required the amputation of all the fingers in his left hand to stop the spread of cancer and he later lost his right hand to ulcers. Both Albert Geyser and his son Frank Roebling Geyser, a practising medical physician, conducted numerous X-ray treatments to establish a method for removing hair with the Cornell tube without causing dermatitis. The method was established through trial and error with no scientific basis. Frank Geyser submitted a patent application for the Cornell tube in 1924 in the US and followed with patent filings in Great Britain, France and Germany. These patents paved the way for further financial investment for the formation of the Tricho Sales Corporation. The company produced and leased the Cornell X-ray machines and trained non-medical operators on the basics of how to use them for the removal of unwanted facial and body hair. Over 75 Tricho beauty parlours opened up across the US, and these salons became one of the most popular places to get such treatments (Figure 4.7).

Advertisements at the time claimed that superfluous hair was beauty's most hideous handicap.

FIGURE 4.7 British advertisement from 1927 for the Tricho System using actress Ann Pennington [1893–1971] as the model. The inference is for the removal of hair from the face, underarms and legs.

Source: British advertisement from 1927 for the Tricho System. Image in the public domain

Modern bathing attire, sleeveless frocks, evening gowns, sheer hosiery and short skirts make it necessary for the gentlewomen to remove the least suggestion of 'Unwanted Hair' on arms, underarms, face, legs and back of neck. Charming features and complexion to rival apple blossom cannot command looks of admiration if marred by hairy growth on lip, chin and body. Such painful embarrassment is unnecessary when science places at your disposal the means for removal of such disfigurement. We use no painful electric needles, powder or corrosive liquids.

The practice of X-ray hair removal persisted in commercial salons for around five decades, long after the scientific community had learned of the carcinogenic dangers of X-rays. The process commonly consisted of a 4-minute exposure to X-rays directly on the face or other skin surface once a week for several months. The treatments generally resulted in the permanent removal of hair but eventually resulted in skin wrinkles, atrophy, patchy fibrous splotches, keratoses, ulcerations and carcinomas. Some women ended up so severely disfigured that they required surgery to remove cancerous growths and tumours. Many women ultimately died prematurely. The term used to describe the characteristic appearance of these unfortunate women was 'The North American Hiroshima Maiden Syndrome' due to the similarity of appearance to the Japanese nuclear bomb survivors. By 1970, one medical report estimated that X-ray hair removal was responsible for over one-third of all radiation-induced cancer in women. The long delay between the radiation exposure years earlier and the subsequent development of cancer observed in the 1970s was to be a typical effect of radiation exposure on the body. Tricho hair salons and spas still exist today and can be found in six locations in the US, although the company states that its 'roots' only go back to 2003.

4.5 MASS IRRADIATION FOR HEALTH AND HYGIENE

In the same way that the use of X-rays for beauty treatments had undesirable effects, mass X-ray epilation treatment carried out for improving health and hygiene resulted in equally bad consequences for many thousands of people. Ringworm, also known as dermatophytosis, is not in fact a worm as the name suggests but a fungal infection that can affect the body and scalp of both humans and animals. Ringworm is caused by three different types of fungus and can occur after contact with soil or an infected individual. The infection produces patches with a characteristic round shape resembling a circular worm, hence the name. Ringworm of the groin inner thighs and buttocks is known as 'jock itch'. Athlete's foot is the common name for ringworm of the foot, which commonly occurs in people who go barefoot in gym showers, locker rooms and swimming pools. Ringworm of the scalp (tinea capitis) is a more disfiguring condition that was particularly common in children up until the 1960s, causing unsightly itchy, scaling bald patches. Ringworm was commonly associated with poor diet, poverty and neglect. Visible ringworm infections were viewed with disgust. The afflicted were considered unclean and stigmatised as people feared infection. Although it was never a reportable disease, in Russia, France and some countries of Central and Southern Europe, the incidence of ringworm was so severe that special schools were established for children with infectious scalps. As mentioned in a previous chapter thallium salts had previously been used to treat ringworm. In addition to manual removal of hair the topical treatments for curing ringworm included carbolic acid, sulphur, wood tar and mercuric chloride that caused painful burning to both the affected areas and the surrounding normal skin, often resulting in further infection. The answer for an inexpensive and effective cure was found with X-ray irradiation. It is thought that two doctors in Vienna, Freund and Schiff, were the first to try X-rays to treat ringworm cases but it was Raimond Sabouraud, a French dermatologist, who gained the reputation for pioneering the X-ray treatment of infected scalps. This method had been systematically developed to reduce the radiological damage to skin whilst delivering a course of irradiation to cause epilation. X-rays had two advantages over the use of fungicides in that they reduced treatment times from years to months and produced a permanent cure.

In Britain ringworm was first identified as a problem after the introduction of the Education Act of 1870 and the start of mass schooling. Inspection of schools revealed the 'verminous condition' of many children and so 'nurseries of ringworm' were created to reduce the spread of infection from classrooms and playgrounds. Policies for ringworm were developed along similar lines to those for diphtheria and scarlet fever. In 1897 London's Metropolitan Asylums Board took specific measures to establish special institutions 'to eradicate the physical taints of pauperism and to place them on a fairer level of health for the race of life'. The first ringworm school was the Bridge School in Witham, Essex, which started in 1901. In February 1903 this was replaced by the Downs Ringworm School (also known as Banstead Road School) in Sutton, Surrey. Children were accommodated in blocks of 70 beds, attended lessons and were given a daily treatment routine of scalp bathing, intensive applications of lotions and the extraction of diseased hairs. Following reports of Sabourad's success the first large-scale use of X-ray treatment in Britain was at the ringworm schools. The success of this treatment led to the closure of the Bridge School at Witham in 1908, saving the asylums board some £500 per year. Treatment was directed by Thomas Colcott Fox, together with the school's medical officer Dr Sale. Between them they developed facilities for multiple simultaneous treatments and carried out a large trial on inmates who had no choice but to comply with the strict discipline. At the time, the cure rates achieved were applauded; however, the ethical and moral standard of such trials carried out with no controls were questionable. The London County Council's Board of Education was so impressed with the results that in 1907 it considered a scheme to provide free X-ray treatment for the capital's children at hospitals and special centres (Figure 4.8).

Despite a growing awareness of the hazards associated with exposure to X-rays similar treatment units were set up in other cities. In March 1909 a letter written by Dr Dawson Turner, who worked in the Electrical Department at the Edinburgh Royal Infirmary, was published in *The Times* newspaper. Turner described himself as an 'old worker with X-rays', who had suffered injury from radiation exposure, and wrote

> The deleterious effects of continuous exposures to X-rays in the case of adults are only too well known to X-ray operators and it is probable that delicate cells of the growing brain of a child may be injuriously affected by much short exposures, though the evidence of impairment of function may not become noticeable until development is complete. No helpless child should have the chief centre of its nervous system exposed

FIGURE 4.8 X-ray apparatus at the Royal London Hospital 1905. This equipment was used for multiple simultaneous treatment of ringworm of the scalp.

to the X-rays without the express consent of its parent, obtained after the possible risks of the treatment have been explained.

Despite this, the use of X-rays continued over the next 40 years. This was not the case in the US where dermatologists argued that oral thallium acetate was safer than X-rays. This turned into a debate over a medical preference for the use of either chemical poisoning or radiation poisoning.

In 1918, after the First World War and the Russian revolution, the old empires had fallen and new countries were established. The Jewish communities, in particular, found themselves under increasing economic hardships and social distress, a situation that continued for over a quarter of a century. Displaced populations faced increased immigration restrictions in many Western countries. Health was a key issue and ringworm was a seen as a significant problem. Records from the US immigration offices on Ellis Island, New York, describe a high incidence of ringworm among immigrant children. People with severe cases were not allowed to enter the US and were returned back across the Atlantic to their country of origin in Europe. In November 1914, Jewish communities in the US founded the American-based Joint Distribution Committee (JDC) as a vehicle for assisting Jewish refugees in Europe during World War I. Its main aim was to give both material and spiritual assistance to the displaced populations and to help Jews to emigrate elsewhere. The American JDC combined efforts with the German- and Polish-based Society for the Protection of Jewish Health (OZE), with a primary goal of promoting good health as a prerequisite for immigration. As part of the 'hygienic preparation of would-be emigrants' a campaign was initiated for the eradication of tuberculosis (bacterial infection of the lungs), trachoma (chlamydia infection) and ringworm. The treatment methods were set up to meet the international medical standards of the time. The treatment of ringworm was aimed at Russian refugee children in particular and the 'state-of-the-art' method of choice was radiation epilation. In 1922 X-ray machines were obtained from the Siemens plant in Erlangen, Germany, and a network of ringworm radiation centres was established in major cities, each headed by a dermatologist or radiologist. Each facility comprised a clinic, a bacteriological laboratory, an X-ray suite, auxiliary services including central bathing facilities, laundry and sanitation services and living quarters for children undergoing treatment. The children with ringworm were identified and admitted to the centres. After microscopic assessment a course of X-ray treatments was administered, followed by the manual removal of any remaining hair (by plucking) and the application of medicinal substances to the scalp. The documentation of each patient treatment was meticulous and following discharge the staff conducted a home visit to assess the domestic setting and instruct the family on home hygiene.

Between the years 1921 and 1938, this mass treatment programme to eradicate ringworm among the Jewish community in Eastern Europe irradiated some 27,600 children with X-rays. This was the largest epidemiological campaign of the early 20th century and was largely successful in eradicating scalp ringworm. Whilst this was carried out with the best of intentions, it could be argued retrospectively that parents sacrificed their children for the sake of the family well-being and security. Curing this infection enabled families to be fit for emigration but the medical irradiation procedure had serious long-term consequences for the children who were individually irradiated. By keeping the treatment radiation exposures at levels below those known to produce erythema and skin burns the radiologists considered the process to be safe. In reality it took decades before the true effects of radiation exposure were appreciated, as evident from the follow up of the Japanese survivors of the atomic bombings at Hiroshima and Nagasaki. Tragically many of the Jewish children irradiated for ringworm in Eastern Europe subsequently perished in the Holocaust. If more children had survived, it is highly likely that the delayed effects of these radiation exposures would have come to light a lot earlier.

Despite the growing awareness of the long-term effects of X-ray exposure and the potential for malignancy, ringworm irradiation remained the treatment of choice throughout the world. In the 1950s Yugoslavia had one of the largest ringworm irradiation programs, with 24 treatment centres. Between January 1946 and December 1960, during the height mass immigration into the newly established state of Israel, 20,000 children were exposed to radiation for ringworm infection with financial backing and equipment from UNICEF. The consequence of these mass irradiations to the head was an increase in the development of malignancies, particularly to the radiation-sensitive tissues such as the brain, thyroid gland and bone marrow. From retrospective studies the estimated radiation dose to a child's brain was in the order of 1.4 Gy, resulting in an increase incidence of brain tumours, including meningiomas and acoustic neuromas. Radiation doses to the thyroid gland in the neck were in the order of 60 mGy, resulting in the production of thyroid adenomas.

The dose to cranial bone marrow has been estimated at around 4 Gy, which led to a reported increase in cases of leukaemia.

This practice of scalp irradiation ended after 1959 when a newly discovered pharmaceutical, griseofulvin, proved to be completely efficacious in the treatment of ringworm and far easier to administer than irradiation. In total it has been estimated that as many as 200,000 children worldwide were exposed to X-rays for the treatment of their fungal infections. This remains the largest intentional irradiation of any population in history.

5 ☢

Kill or Cure

Nothing in life is to be feared,
it is only to be understood.
Marie Curie

5.1 A STATE OF DECAY

Radioactivity surrounds us. It is a natural part of the world we live in, has been integral in the geological formation of the earth and has played a role in the evolution of all life. Some scientists will argue that radiation is the same as other physical entities in the world, such as heat and water, in that small amounts are essential to life, whereas too much will kill you. The majority of naturally occurring radioactive materials are found in rocks of the earth, but radiation is also in the sky above us, in the buildings we live in and in the food we eat. In fact, we are all radioactive! Some of life's essential elements which are vital for cellular function, such as carbon which has a role in biological pathways and potassium which is an important blood element, have natural radioisotopes that are taken up by body tissues.

Radioactive materials are described as being unstable and have the natural ability to change the configuration of the particles in the nucleus of their atoms. As described in Chapter 2, the atomic nucleus of all elements contains two fundamental types of particles known as protons and the neutron. The protons have a positive electrical charge and will force each other apart (in the same way that the same poles of a magnet repel each other) if it were not for the presence of the neutrons which are neutral, i.e. have no charge. The neutrons therefore weaken the repulsive force of the protons allowing the nucleus to remain stable. For most stable atoms the number of protons is more or less equal to the number of neutrons. However, if the number of protons and neutrons diverge the atom is more unstable and will correct the imbalance by a process known as radioactive decay. If the atom has too many protons (proton excess) a proton will be converted to a neutron, whereas if there are too many neutrons a neutron is converted into a proton. These nuclear processes take place to allow a more stable balance of protons and neutrons and may be repeated through a number of transformations known as a decay chain. A fundamental feature of these transformations is that the conversion of protons into neutrons and neutrons into protons also releases other forms of energy from the nucleus. This energy is mainly in the form of radiation and is mainly of three types: alpha, beta and gamma. In some situation neutrons may be emitted, but this is mostly associated with nuclear fission or fusion reactions and alpha particles can cause neutrons to be emitted when they interact with the atoms in other materials.

5.2 ALPHA, BETA AND GAMMA RADIATION

Unstable atoms become more stable by giving off radiation which is in two basic physical forms: particle radiation and electromagnetic radiation. Figure 5.1 illustrates some of the properties of these types of radiation.

Alpha particles are the heaviest and largest form of particle radiation containing two neutrons and two protons. This is the same as the nucleus of the helium atom and so has a positive charge. Being large they do not penetrate far into matter and can be stopped by a thin sheet of paper. However, alpha particles can deposit a large amount of energy in a small volume of material. If this energy is absorbed by tissues of the body it can cause a great deal of biological damage, for example by cutting strands of DNA in cellular chromosomes or by producing chemically active free radicals in tissues, both of which can cause tumours to develop.

Some atoms with an excess of neutrons may attempt to reach stability by converting a neutron into a proton with the emission of an electron. The electron is called a beta-minus particle, the minus indicating that the particle is negatively charged. If the number of protons in the nucleus is too large for the nucleus to be stable, it may move to a more stable state by converting a proton into a neutron with the emission of a positively charged electron which is called a beta-plus particle or positron. The Italian physicist, Enrico Fermi, showed that these positively charged electrons were

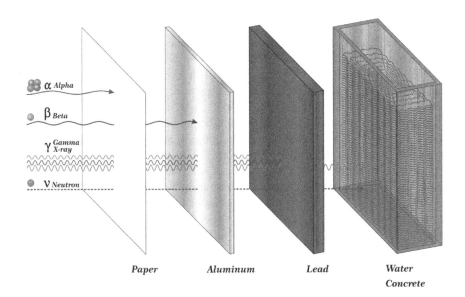

FIGURE 5.1 Diagram showing the absorption of alpha, beta, gamma, X-rays and neutrons in common materials.

Source: Diagram showing the absorption of alpha, beta, gamma, X-rays and neutrons in common materials. www.shutterstock.com/image-vector/alpha-beta-gamma-neutron-radiation-penetration-1321327076

in fact antimatter that does not exist for very long, as they rapidly combine with a normal electron in a reaction called annihilation. This gives rise to the emission of two gamma rays that are emitted travelling in opposite directions. This process is used in modern-day medical imaging, in a technique known as PET scanning (positron emission tomography). Beta particles can travel many centimetres through air but are absorbed by small thicknesses of light materials such as aluminium, glass or acrylic plastics.

Gamma decay involves the emission of energy from an unstable nucleus in the form of electromagnetic radiation similar to light, radio waves, infrared, ultraviolet and X-rays. The properties of gamma and X-rays are the same, the only difference being that X-rays are produced by the electrons surrounding the atom while gamma rays emerge from within the nucleus of the atom. Gamma and X-rays are the most penetrating forms of radiation capable of passing many metres through air and solid material. Dense absorbing materials such as lead or concrete are needed to attenuate these rays and provide protection against human irradiation.

5.3 RADIOACTIVITY AND THE BIRTH OF THE ATOMIC AGE

Natural radioactivity was nearly discovered in 1857 when a French photographic researcher Claude Félix Abel Niepce de Saint-Victor was working under the chemist Michel Eugène Chevreul. Whilst using light-sensitive metal salts to make colour photographs Chevreul realised that uranium salts emitted something invisible to the human eye that would expose photographic emulsions. He observed this effect in complete darkness and also demonstrated that it could be detected on photographic plates months after the metal salts had any previous exposure to sunlight. Chevreul recognised this as a fundamental discovery, but it is the French physicist Antoine Henri Becquerel who is credited with the discovery of radioactivity, although he did not name it as such at the time. Becquerel came from wealthy parents and was the third in his family to hold the chair in physics at the National Museum of Natural History (Muséum National d'histoire Naturelle) in Paris. After completing his PhD on the polarisation of light Becquerel continued to work on the fluorescent properties of chemical substances that could store and release visible light. In 1896 Becquerel was working on the theory that exposing some natural minerals to sunlight gave rise to the production of something similar to X-rays. He placed an array of fluorescent chemicals on a sheet of photographic film wrapped in black paper, allowing the sunlight to energise the substances to determine any fluorescence effects. The only material which produced any positive result was uranium sulphate, so he carried out further experiments with uranium. On one particularly overcast day in November Becquerel abandoned his experiments and placed the uranium and film packet in his desk draw. The next day he processed the film to discover an intense image of the granules on the film, showing that it had been exposed without activation by sunlight. He then realised his mistake of previously failing to include a control test to show whether it was the sunlight or the uranium itself that was producing the effect. He finally concluded that the uranium was spontaneously emitting some form of invisible radiation similar to X-rays. He named the new radiation phenomena 'uranium rays' but they were often referred to as Becquerel rays. Since the images of uranium rays did not show the same degree of detail as Roentgen's X-rays, which were capable of taking pictures of the human body, Becquerel's scientific discoveries did not receive widespread interest from the general public. The term 'radioactive' had not yet been coined but this was the start of a road of scientific discovery that would change the course of history. Radioactivity was the name subsequently used by Marie and Pierre Curie to describe the spontaneous emission of radiation from unstable elements. In 1903 Henri Becquerel was awarded the Nobel Prize in Physics, together with Marie and Pierre Curie, who went on to carry out more detailed investigations into the nature of uranium ore.

Marie Curie's story is well known. It is one of a naturally shy and timid person who through persistence and hard work rose to the highest intellectual heights of her time. Her story is all the more remarkable given the male-dominated culture of academia in the late 1800 and early 1900s. Maria Salomea Sklodowska was born in Warsaw in 1867, a time when Russia was grinding down hard on Poland, a country that had previously been overrun by Swedish, Muscovite, Austrian, Tsarist and Prussian armies and would later be invaded by Nazi Stormtroopers. When she was 10 years old, Marie's mother died of tuberculosis. As a child Marie was sent to a boarding school and then attended a gymnasium for girls where she received a gold medal. Further education was disrupted due to a breakdown and she spent a few years living with relatives in the country where she undertook some tutoring work. Being a woman, she was unable to study at any national higher educational establishments but together with her sister managed to enrol with the Polish Flying University, or 'Floating University', an institution that frequently changed its location to avoid detection by the Russians. The Floating University taught physics and natural history as well as forbidden subjects such as Polish history, but more importantly for Marie and her sister it would admit women students. In 1891 she left Poland to live in Paris with her sister and brother-in-law and later moved into a garret close to the University of Paris, where she was studying physics, chemistry and mathematics. After obtaining a degree in Physics she took a job at an industrial laboratory whilst continuing with a fellowship at the university to get her second degree. Whilst investigating the magnetic properties of steel she met Pierre Curie, a 35-year-old instructor at the City of Paris Industrial Physics and Chemistry Educational Unit. They became close through their mutual passion for science and were married in 1895 (Figure 5.2).

Pierre had a modest salary from his teaching position and this allowed Marie access to his laboratory where she studied without pay. Together they lived a simple life in an apartment on the rue de la Glacière within a short walk from the laboratory. Marie was intrigued by the recent discoveries of Roentgen and Becquerel and decided to work on identifying the nature of the emissions from uranium salts as a possible subject for her doctoral thesis. The Curies considered that uranium ore contained other elements that were more radioactive than uranium. Marie was given space in a small part glazed room that had been a machine shop and storeroom on the ground floor of the School of Physics and Chemistry. The room was cold and damp and there was no funding for experimental apparatus. The one piece of equipment that she had was an electrometer that Pierre had built some years earlier. This device consisted of a brass box containing a thin piece of quartz crystal, sandwiched between two rods which could detect the electrical charge from an ionisation chamber. Placing a test sample on the metal plate allowed a simple and rapid test to see if any electrical charge

FIGURE 5.2   Photograph of Pierre and Marie Curie taken around 1900.

Source: Photograph of Pere and Marie Curie taken around 1900. www.shutterstock.com/image-photo/
pierre-curie-marie-sklodowska-18671934-c-339956723

was produced from the material. Marie made an initial discovery within days of starting work when she observed that uranium and thorium samples both produced a current. Further observations showed that larger samples produced more current and the
same effect was observed in wet and dry solid and powder forms. Marie had observed
the production of energy from this material without any chemical transformation or
visible change of form taking place. These observations were in conflict with the
accepted basic law of thermodynamics concerning the conservation of energy. She
went on to propose that this may be a basic property of the uranium atoms themselves
which were releasing energy. Her suggestion that subatomic particles existed was
something that challenged the long-held scientific belief that atoms were indivisible.

Pierre put his own work aside to help Marie with her experiments. They started
with several tonnes of a black tar-like substance called pitchblende, which came
from mines in Bohemia. The name 'pitchblende' is derived from the German word
'pech' for the black colour and 'blenden' from the German word for 'blind' or 'daz-
zle' which was due to the complex mixture of heavy ores. Since Roman times the

ores in pitchblende were used to add a yellow colouring to glass, pottery and ceramic tiles. In the early 1780s Martin Klaproth, a chemist who later became Professor of Chemistry at Berlin's Royal Mining Academy discovered that this black mineral could be used to give glass an intense yellow colour. He was certain that it contained a new metal which he named uranium after William Herschel's discovery of Uranus in 1781, which was named Uranus after the Greek god of the sky. Today pitchblende is known as uraninite and is known to contain mainly uranium oxide (U_3O_8) and its decay products. In the 1930s uranium was added to glass prior to melting to produce a vivid yellow/green colour which was enhanced by lighting. This was used in a range of glass ornaments, vases and table ware commonly known as Vaseline glass due to its similarity in appearance to Vaseline petroleum jelly which was produced at the time. Today the atomic energy industry uses uraninite to produce powdered yellowcake (urania), the starting material for further enrichment of the materials used for nuclear fuel and in nuclear weapons. It is mined in the McArthur River Mine Canada, the Olympic Dam Mine, north of Adelaide and the Ranger Mine east of Darwin, Australia, the Arlit Mine in Niger, West Africa, Rossling in the Erongo region of Namibia, at Budenovskoye 2 and South Inkai in Kazakhstan and Kraznokamensk in Russia near the Sino-Russian border.

The pitchblende used by the Curies was delivered in carts and deposited in the yard next to the shed where they were working. They set about with shovels dissolving the ore in cauldrons of acid and extracting salts for conversion into chlorides. The process required over a thousand crystallisations with each incremental step guided by the electrometer readings, as the pure elements containing higher readings were isolated. The poor working conditions in the ramshackle accommodation with widely varying temperatures and humidity made the taking of experimental measurements difficult. Both Marie and Pierre suffered burns, fatigue and various illnesses as a result of the work and exposure to radiation. Today their laboratory notebooks are still regarded as being too radioactive to handle so they are kept in secure, shielded containment at the Historical Resources Centre of the Curie Museum in Paris.

In 1896 Marie identified the first pure substance and gave it the name 'polonium' after the country of her birth. Uranium ores contain minute traces of polonium of the order of parts per billion. Nevertheless, the Curies had been able to separate out small amounts of natural polonium-209 with a physical half-life of 103 years after months of painstaking work in little more than a garden shed. They had found the element with an atomic number 84 next to bismuth that had been predicted many years earlier in Mendeleev's periodic table. Despite the arduous nature of the separation process, production of natural polonium followed. The alpha particles it gave off had the property of neutralising electrical charge in air making it suitable for use in anti-static devices. Other applications would emerge decades later after the scientific and technological endeavours that would result in the production of artificial radioactivity.

Five months after the discovery of polonium Marie identified a second element which she named 'radium' after the Latin word 'radius' which means 'ray'. Four years of physically exhausting work had resulted in the extraction of 1 milligram or radium from 10 tonnes of pitchblende ore. Radium is an alkaline metal with an atomic number of 88. It has physical and chemical properties similar to those of barium. Pure radium is a volatile silvery white metal which becomes black when exposed to air. The Curies discovered that natural radium-226 was the most stable form of naturally occurring radium, with a physical half-life of 1600 years, yet it is millions of times more radioactive than uranium. Since its discovery a total of 33 isotopes (atoms having the same number of protons and electrons but different numbers of neutrons) of radium have been identified. Marie Curie came up with the word 'radioactive' to describe the property of the elements she discovered. In 1903 she was awarded the Nobel Prize in Physics jointly with her husband Pierre and Henri Becquerel. This award was initially only going to be given to the two men, but Pierre fought to get her name added and she became the first woman to be awarded a Nobel Prize. Pierre was promoted to a full professor at the Sorbonne, although as was typical for the time, Marie being a woman was not given employment. Pierre appointed Marie as the laboratory head and was allowed to continue with her research and, for the first time, get paid for her work. Three years later in 1906 Pierre was suddenly killed as he absent-mindedly stepped in front of a carriage whilst crossing the Rue Dauphine.

Marie was left with two daughters to look after. Irene, the eldest developed an interest in science and worked with her mother before marrying Frederick Joliot, a chemical engineer. The Joliet-Curies went on to calculate the mass of the neutron and were acknowledged for creating the first artificial radioactive material by bombarding non-radioactive elements with alpha particles. They received the 1935 Nobel Prize for chemistry in recognition of these achievements.

The younger daughter, Eve, was not interested in science but was more inclined to literary works and journalism. Between 1941 and 1942 she travelled to Africa, Russia and Asia as a war correspondent.

Marie took over Pierre's position at the Sorbonne, although she met much resistance from the exclusively male scientific establishment. The Curies had taken a deliberate decision not to patent polonium or radium, but after Pierre's untimely death Marie took control of radioisotope production at the Paris laboratories, creating a lucrative income stream for further research. Her ongoing career was not without controversy, when in 1911 rumours emerged that she was having an affair with Paul Langevin, a prominent physicist 5 years younger than her, who had been Pierre's student and had worked with Albert Einstein. Later in the same year she was awarded the Nobel Prize in Chemistry for the discovery of polonium and radium, but by the end of the year she became seriously ill and required surgery to remove tumours from her uterus. Following the outbreak of the First World War, encouraged by her daughter Irene, Marie was instrumental in setting up some 18 mobile X-ray units for use at the front line. Irene completed a nursing course so that she could help her mother as a nursing radiographer on the front line. These units proved invaluable for the localisation of bullets and shrapnel in wounded soldiers and army personnel.

In 1934 at the age of 66, Marie Curie died of aplastic anaemia, a disease in which the body fails to produce sufficient numbers of blood cells. In Marie's case this was due to the to the long-term effect of radiation on her bone marrow. For over 30 years she had been exposed to radioactivity and X-rays, whilst working in Paris and later on the battle fields. She was known to carry test tubes containing radioactivity in her pocket and stored them in her desk drawer. It was said that she liked the blue-green light that the radiation produced in the dark. The cumulative effects of radiation exposure had caused near-blindness as a result of radiation-induced cataracts and chronic illnesses which ultimately led to her death. Throughout her life Marie never really acknowledged the real health risks of exposure to radiation. After Marie's death Eve wrote her mother's biography.

5.4 RADIUM, THE ELEMENT OF LIFE AND DEATH

If X-rays had taken the world by storm, then radioactivity created a whirlwind. The Curies discovery of polonium and radium generated huge interest amongst the scientific community and word spread rapidly through the newspapers and journals of the time. The concept of radioactivity challenged the hitherto accepted scientific knowledge since until this moment in history the very word 'atom' meant 'that which could not be subdivided'. At the time of its discovery, interest in the use of polonium was limited in comparison to the surge in interest resulting from the use of radium which has far more radioactivity. It was subsequently discovered that radium-226 was part of the uranium-238 decay chain that produced a number of radioactive elements that also included radioactive protactinium, thorium, radon, astatine, polonium, bismuth, thallium and mercury before reaching stable lead-206.

Mysterious and intensely powerful, emitting energetic rays that emanated a soothing glow in the dark, element number 88 in the periodic table was regarded as incredible and fascinating. This was an exciting new era of chemistry, when in 1903 the medical journal *Lancet* reported that the discovery of radioactivity was taking science nearer than ever before towards glimpsing at the very nature of things. Many scientists and botanists considered that this half-living element held the key to the secret of life. By 1903 experiments carried out by Ernest Rutherford and Frederick Soddy at McGill University in Montreal had shown that as a result of radioactive decay, elements could transmute from one to another. Radioactive transmutation was a throwback to the realms of alchemy giving rise to the idea that some atoms were alive and that life and the universe had evolved through transmutation of the elements.

Radium was considered to be energising and capable of revitalising both body and soul. In other words, it was *the philosopher's stone*. This incredible new element could do no wrong, and it became regarded as an industrial shining light and a truly modern enhancer of life, health, beauty and growth. This was a time along way ahead of any realisation of the harmful effects from radiation exposure and before any idea that radioactive atoms could be used to make the most powerful weapons on earth. Once the radium separation process had been understood, it could be produced in larger amounts and used for the benefit of everyone. Since only small amounts of radium would produce visible light it had a variety of uses in a number of products, in particular radioluminescent items for night vision including:

Clock and watch hands and face markings
Gauges and dials
Compass needles
Buttons for door handles
Light switches
Military gun sights

Military personnel markers
Exit signs
Chain pulls
'Glowboy' fishing lures

Photographs of the radium ray spotter button and a luminescent voltmeter are shown in Figures 5.3 and 5.4.

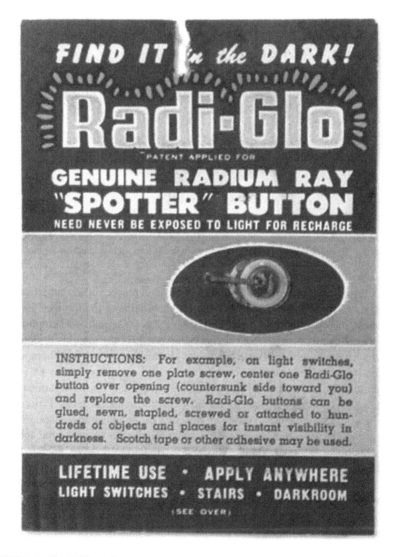

FIGURE 5.3   Radi-Glo radium ray spotter button and instructions for use.

Source: Radio-Glo radium ray spotter button and instructions for use. Printed with permission from the Oak Ridge Associated Universities Health Physics Historical Instrumentation Collection www.orau.org/ptp/museumdirectory.htm

FIGURE 5.4   Luminescent voltmeter gauge.

Source: Luminescent voltmeter gauge. Printed with permission from the Oak Ridge Associated Universities Health Physics Historical Instrumentation Collection www.orau.org/ptp/museumdirectory.htm

The luminescent properties of radium had obvious practical uses but looking back it is even more remarkable how widely the health benefits of radium were embraced. Radium was incorporated into a plethora of products for human consumption, injection, application and insertion into the body. In April 1914 the monthly scientific journal *Radium*, which published articles on the chemistry, physics and therapeutics of radium and radioactive substances, contained a scientific paper written by Dr Frederick Proescher on the intravenous injection of soluble radium salts for the treatment of high blood pressure. Unfortunately, such treatments had no scientific basis! Proescher had treated 34 volunteer patients, 16 of whom had arthritis. Another advocate of the medical use of radium was a physician, Dr C Everett Field, who was also manager of the Radium Chemical Company's New York office. He had a lucrative practice until the late 1920s and enthusiastically published the virtues of radium therapy in medical journals and in his own printed pamphlets. In one publication in 1926, Field wrote that he had administered 6,000 intravenous radium treatments over a 12-year period, although there is no record of any long-term follow-up concerning any effects of these treatments.

By the 1920s it was possible to purchase drinks, tablets, gels, ointments, beauty creams, hair products, toothpaste, ear plugs, suppositories, contraceptives, bread, chocolate bars and even cigarettes, all containing radium. Radium soap and radium hand cleaner were advertised to take off everything but the skin! In France, the company Tho-Radia sold a radium beauty cream with a formula alleged to have been produced by a Doctor Alfred Curie, who in fact never existed. The cream contained

0.25 millionth of a gram radium bromide in each 100 grams, a formulation that the company claimed would eliminate wrinkles from the face under the premise that science has created Tho-Radia to beautify women (Figure 5.5).

Soon after Marie Curie visited America in 1921 the sale of radium products soared. The Associated Radium Chemists company in New York sold radium health tablets under the name 'Arium'. Vita Radium Suppositories were sold for rectal use by men, for restoring sex power and energising the entire nervous, glandular and

FIGURE 5.5   French poster for Tho-Radia beauty creams, circa 1933.

Source: French poster for Tho-Radia beauty creams, circa 1933. Printed with permission from the Oak Ridge Associated Universities Health Physics Historical Instrumentation Collection. www.orau.org/ptp/collection/quackcures/Tho%20Radia%20Cards.html

circulatory system. These contained refined soluble radium in a cocoa butter base and were designed to allow absorption of radium through the rectum and colon into the bloodstream to allow circulation to the affected area. These were highly recommended for 'sexually weak' or impotent men and were most effective when taken together with 'Nu-Man' tablets. An end note stated that they were also splendid for piles and rectal sores. The words *atomic piles* come to mind, but there was no mention of this as a possible side effect!

The American Endocrine Laboratories produced a device called the Radiendocrinator, which was a small pad with a metallic surround containing radium to be used for placing on parts of the body to 'radiate as directed'. The instructions stated that it should be worn as an athletic strap, placing the device under the scrotum to aid male virility, allowing men to bubble up with joyous vitality. Another appliance was the Testone Radium Energiser and Suspensory which contained 20 micrograms of refined, measured radium. The sellers claimed that this marvellous appliance would give new energy to weak sagging men and had no equal for shearing up and husbanding strength in that area of all male activity – the testicles.

5.5 THE QRAY ELECTRO-RADIOACTIVE DRY COMPRESS

In July 1988 two of my colleagues in Medical Physics at Queens' Medical Centre, Nottingham, were called out by the Nottingham police to provide assistance in recovering a radioactive artefact that had come to light in the county.[2] A science teacher from Brunts School in Mansfield had purchased a 40-year-old electric blanket at a church bazar. He was intrigued after reading the label on the box that stated that the blanket gave off both heat and radiation. He took it to his teaching lab at the school and on testing it with a Geiger counter confirmed that it gave off radiation. The school headmaster was alarmed when he was told that radioactive material had been brought into the school. He was concerned that this could expose young students to harmful radiation and called the police. The device was a Qray radioactive electric compress that was sold for use in hospitals and in the home in the 1930s and 1940s by Radium Electric Ltd in 153 Victoria Street, London. A photograph of the appliance is shown in Figure 5.6.

At the time of recovery it was estimated that the compress contained 0.5 MBq radium-226 and the surface dose rate was around 5 micrograys per hour, which did not represent any danger to those who came near it. The product information showed that it was designed to combine the curative properties of radioactivity with heat to 'allay pain, assist circulation and promote healing' (Figure 5.7). It was stated that there was abundant information from hospitals and the medical profession on the functional restoration and treatment of rheumatism, lumbago, arthritis, gout and

---

[2] On 2 July 1986 Dr John Hardy (Medical Physicist) and Dr Malcom Frier (Radiopharmacist) responded to a request from the Nottingham Constabulary under the UK National Arrangements for Incidents involving Radioactivity (NAIR scheme). They accompanied Superintendent Tordoff during a visit to Brunts School, Mansfield, and safely removed the Qray compress after confirming that it was intact and had not caused any radioactive contamination. The compress was collected the following day for disposal by the National Radiological Protection Board.

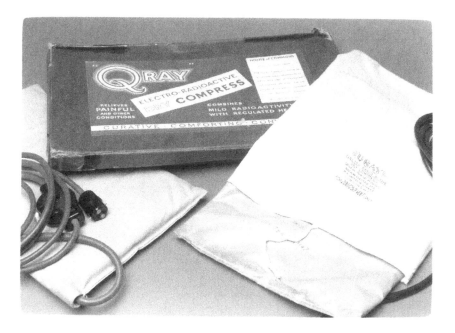

FIGURE 5.6   The Qray electro-radioactive dry compress.

Source: Qray blanket. Author's original material

sciatica as well as bronchial complaints, pleurisy, pneumonia, emphysema and affections of the ear and sinuses[3].

There was also a report on the radioactivity in the pad, which was carried out by the National Physical Laboratory in Teddington (Figure 5.8) and photographs of the compress in use at St Thomas's Hospital London (Figure 5.9). The information also contained an article form the *Chelsea Chiropodist Journal*, dated August 1939, on the uses of mild radioactivity in chiropody, an extract from the 1937 Annual General Meeting of the National Union of Railwaymen's Approved Society describing an amazing successful experiment of 'Qray' treatment of rheumatism and kindred inflammatory complaints and the records of the Ministry of Health debate on 8 June 1937 taken from *Hansard* on the merits of the Qray compress that were brought before the House of Commons. In hindsight, some patients may have experienced some therapeutic benefit from the heat produced by the blanket and the amount of radioactivity would not have caused serious harm unless it was used excessively.

5.6 SUNSHINE IN A BOTTLE

Of all the radium medical products the radium drinks proved to be the most popular and also one of the most harmful products. The idea that natural spring

---

[3] At the time, 'affections of the ear' was a term that was used to include a range of conditions affecting the ear canal, including loss of hearing, infection, inflammation and discharge. This also included conditions resulting from venereal diseases such as syphilis, which can cause hearing loss.

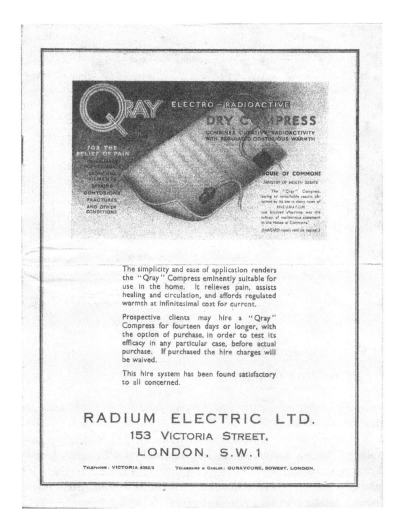

FIGURE 5.7   Qray compress product information.

Source: Qray compress information. Author's original material

waters contain health-giving properties goes back centuries. Throughout history people would travel long distances in search of healing waters. In 1903, scientist J. J. Thompson, who discovered the electron, wrote a letter to the journal *Nature* in which he described the presence of natural radioactivity in well water. This led to the findings that the waters of many world-famous health spas contained varying amounts of radioactivity. For this and other reasons spas and baths became fashionable health resorts in the 20th century. These were places like Bath and Buxton in England (both natural spas, dating back to Roman times), Bad Gastein in Austria and Bad Berka in Germany. In Canada you can still visit the village of Radium Hot Springs on the edge of the Rocky Mountain National Park, where you can walk along Radium Boulevard, visit the Radium Liquor Store and swim in the open air

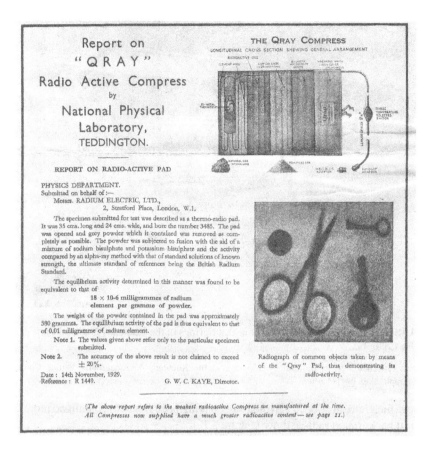

FIGURE 5.8   Report from the UK National Physics Laboratory in Teddington, dated 1929.

Source: Qray NPL report. Author's original material

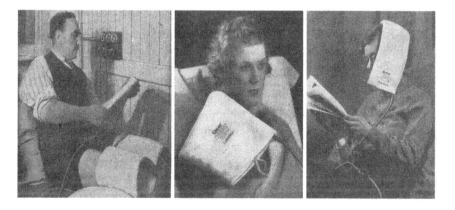

FIGURE 5.9   Patients being treated with the Qray compress. Left: treatment of the knees. Middle: treatment of the shoulder. Right: treatment of the ear.

Source: Qray Patients. Author's original material

FIGURE 5.10   Entrance sign to Radium Hot Springs in the Kootenay National Park, British Columbia.

Source: Entrance sign to Radium Hot Springs in the Kootenay National Park, British Columbia. Photograph taken by the author

pools which have now been certified safe, having been found to contain no more than natural background radioactivity (Figure 5.10).

In the early 1900s 'taking the waters' was considered to be of far greater benefit to health if it contained radioactivity. Professor Bertram Boltwood of Yale University explained that radioactivity was carrying electrical energy into the depths of the body, subjecting the juices, protoplasm and nuclei of the cells to an immediate bombardment by explosions of electrical atoms. This stimulated cell activity, aroused all secretory and excretory organs causing the system to throw off waste products and destroy bacteria. Whilst developing commercial products a key finding was that radium decayed by giving off radon gas which was lost through radioactive decay and escape of the gas to the air during transit from the production site. The marketing solution was to manufacture a jar containing radium ore that was filled with several litres of water (Figure 5.11).

One particular vessel known as the Radium Revigator was widely sold in the US. The radium ore would give off radon into the water which could be freely drunk at home both morning and night. These domestic perpetual health springs could be placed on a sideboard at home and smaller devices known as radium emanators could be carried around in a pocket or bag and placed in a jug of water for use during travel. A radium emanator came into my possession after a colleague[4] picked one up at a jumble sale near Nottingham. It was the Radigam Radioactive Emanator sold by

---

[4] I am grateful to Simon Lawes, Senior Technologist and Radiation Protection Supervisor in Nuclear Medicine at Nottingham University Hospitals NHS Trust, who gave me the vintage Radigam emanator after discovering it amongst various household items at a local jumble sale in Nottingham.

FIGURE 5.11 Radium water preparations available for use in the home in the early 1900s.

Source: Radium water preparations available in the early 1900s. Printed with permission from the Oak Ridge Associated Universities Health Physics Historical Instrumentation Collection www.orau.org/ptp/museumdirectory.htm

the Radiopathic Institute Limited of London in the 1930s. This was a hollow ebonite (bakelite) rod which contained a small amount of radium. A series of fine holes in the rod allowed the water to enter the tube and the radioactive gas to escape producing 'activated water' (Figure 5.12).

The fact that most users could not measure the radioactivity in these products led to the sale of many bogus products. From 1916 to 1929 the American Medical Association specified that a standard activity of 74,000 Bq (2 µCi) of radon should be

FIGURE 5.12   Radigam radium emanator and leather carrying case originally from Radium Spa, London.

Source: Radigam emanator. Photograph taken by the author

produced per litre of water in a 24-hour period for approved devices. In hindsight it is ironical that those selling the safer product (containing no radioactivity) were the ones prosecuted for fraud.

One particular health tonic, *Radithor*, also known as Perpetual Sunshine, was widely prescribed by physicians and practitioners (Figure 5.13). The product packaging described this as being just a tiny bottle of apparently lifeless, colourless and tasteless water which was all that the eyes could see or the tongue could detect, yet it contained the greatest therapeutic force known to mankind. The manufacturers claimed that as well as being an excellent energy tonic, it had proved to be highly valuable for the treatment of conditions including anaemia, arthritis, diabetes, high blood pressure, menstrual disorders, nervous conditions, obesity, senility and sexual conditions.

In the US Radithor was produced and distributed by the Bailey Radium Laboratories which was located at 336 Main St, East Orange, New Jersey. The owner of the company, William J. A. Bailey, had previously been convicted of fraud concerning false claims for a rather dubious cure for impotence called 'Las-I-Go for Superb Manhood'. Despite having no academic qualifications, he restarted his business venture in 1922 setting up radium laboratories including the Association of Radium Chemists in New York, where he manufactured the Arium radium tablets. These tablets promised to restore happiness and youthful thrill into the lives of married people whose attraction to each other had weakened. The Bailey Radium

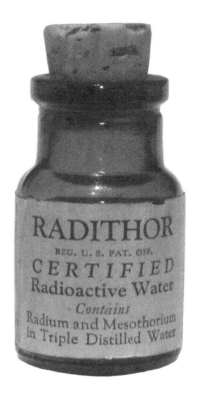

FIGURE 5.13   Radithor health drink

Source: Radithor health drink. Printed with permission from the Oak Ridge Associated Universities
Health Physics Historical Instrumentation Collection www.orau.org/ptp/museumdirectory.htm

Laboratories in East Orange was his most financially successful venture. These were
in operation between 1925 and 1931 when a Federal Trade Commission order ruled
that they should cease from making the therapeutic claims of the product. Radithor
was sold in cases containing 30 bottles (1 month's supply) for $30. Each 2-oz bottle
was claimed to contain triple distilled water guaranteed to contain at least 1 micro-
curie (37,000 Bq) each of the radium-226 and radium-228. Bailey offered $1,000 to
anyone who could prove the product contained less than the stated amount of radium.
No one ever challenged him on this. The company was forced to close in 1931 after
the personal testimony of Ebenezer McBurney Byers, who had been the chairman
of AM Byers Co in Pittsburgh, one of the world's largest iron and steel companies
at that time. Byers had taken over the company founded by his father who was a
wealthy industrialist. After suffering an injury, he had been an enthusiastic propo-
nent of the health-giving benefits of Radithor, but at the age of 51 he was in a bad way
and literally falling apart. In his youth Byers had attended Yale College where he was
known for his athletic prowess and an over-active libido. In 1902 he became the US
Amateur golf champion and also won a number of shooting trophies (Figure 5.14).

FIGURE 5.14   Ebden Byers. Six-time WPGA Amateur Champion and one of the best amateur players in the national amateur's history.

Source: Ebden Byers. Six-time WPGA Amateur Champion and one of the best amateur players in the national amateur's history. Courtesy of the Western Pennsylvania Golf Hall of Fame with permission of the Western Pennsylvania Golf Association

Widely known for his wealthy lifestyle, he had a private box at Forbes Baseball Field, Pittsburgh, and owned racing stables in England and the US. He had homes at Pittsburgh, Southampton on Long Island and Aiken, South Carolina, and he often visited Palm Beach. In 1927 at the age of 51 Byers was returning by train from a Yale-Harvard football game when he fell from the upper berth of a sleeping car and injured his arm. He consulted Charles Moyer, his doctor in Pittsburgh, complaining of severe pain. Moyer prescribed 'sunshine in a bottle' no doubt incentivised by the 17% rebate offered by Bailey Labs on every prescribed bottle of Radithor. Whether it was the placebo effect or otherwise, Byers was convinced that Radithor was having invigorating and beneficial effects. From December 1927, he drank an average of three bottles a day over the following 2 years. This amounted to over three times the lethal dose of radium. Feeling invigorated by his medication he enthusiastically advocated the benefits of Radithor to everyone he met. He gave cases of it to his business colleagues, fed it to his racehorses and he made sure that his girlfriends had plenty to drink. As a young man at Yale, he had not been known as 'Foxy Grandpa' for no reason! Unfortunately, the perceived beneficial effects of this remedy did not last long. He stopped drinking Radithor in 1930 when his teeth started falling out and one of his girlfriends died suddenly of mysterious causes. His lower jaw started crumbling and had to be removed surgically. Shortly after his upper jaw had to be removed. He developed lesions in his brain and his skull began to split open. In the end he had to wrap his head in bandages to hold it together. It was because of Byers position in society and his celebrity status that this horrific illness led to the Federal Trade Commission order that was filed against the Bailey Radium Laboratories. When the Trade Commission were conducting interviews on the efficacy of the product, Byers was too ill to attend the hearing. The examining attorney Robert Winn visited Byers at his grand home on Long Island, New York, and was horrified with what he saw. His report stated that a more gruesome experience in a more gorgeous setting would be hard to imagine. Byers entire lower jaw and chin were missing. Two chipmunk teeth protruded from a fragment of bone below his nose and he had cracks in his skull exposing his brain. After Byers died in 1932 the autopsy report recorded the cause of death as necrosis of the jaw, abscess of the brain, secondary anaemia and pneumonia. Subsequent findings estimated that he must have consumed around 1,400 bottles of Radithor and that his body contained 2,738 kBq (74 microcuries) of radium. He was buried in a lead-lined coffin in the Byer Mausoleum at the Pittsburgh Alleghany Cemetery. His body was later exhumed and measurements showed that he had consumed a total of 12,913 kBq (349 microcuries) of Ra-226 and 22,200 kBq (600 microcuries) of Ra-228. This was considered to be over three times the lethal dose of radium.

After the closure of the radium laboratories in 1937 William Bailey went on to become a partner in the Lee Kelpodine Company of New York City. The sole product line consisted of Kelpodine Tablets made of compressed pelletised seaweed. The company's claims that the tablets would alleviate 32 specific diseases and other conditions caused Bailey to come in conflict with the US Food and Drug Administration. William Bailey died of bladder cancer at the age of 64. His body was later exhumed in 1970, and it was found to contain significant amounts of Ra-226 and Ra-228.

5.7 THE POISONED PAINTBRUSH

As mentioned previously, Marie Curie was known to enjoy looking at the strange glow from radium. In 1906 Sabin von Sochocky emigrated from Europe to the US and developed a luminous paint formula containing radium. After selling 200 watches with luminescent dials he formed a company to finance his medical studies. His radioluminous paint consisted of radium and a phosphor compound. The radiation emitted as the radium decays excites the phosphor producing visible light. Together with George

Willis he formed The Radium Luminous Materials Company, which was established in Newark, New Jersey, in 1915, but the company moved to larger premises in Orange, New Jersey, in 1917 and became the US Radium Corporation. The luminous paint called 'Undark' was seen to be of value to the US military and the company became a major supplier of watches and dials used by soldiers during the first word war Figure 5.15.

FIGURE 5.15 Magazine advertisement for Undark produced by the Radium Luminous Materials Corporation in 1921.

Source: 1921 advertisement for Undark produced by the Radium Luminous Materials Corporation. Image in the public domain

The company profited and a radium extraction plant and a dial painting studio was established. The dial painting work was easy and well paid. It was almost exclusively for young women and occasionally disabled people, since the work was mainly carried out sitting down. Between 1917 and 1926 the US Radium Corporation hired a total of 4,000 employees. The job was to paint dials, clock faces and watches with radium paint (Figure 5.16).

This required a fine camelhair brush to produce thin painted outlines on the clock face or dial, a job that young women with small steady hands were considered to be well suited to. The girls were trained in the technique of lip pointing. Before each dip of the brush in the paint they would carefully place the brush tip between their lips to form a fine point. The gritty paint had a slightly bitter taste but they were reassured that there were no harmful effects, after all radium was good for health! This *lip and dip* process was repeated throughout each working shift and with each lip pointing a small amount of radium was ingested. The metabolism of radium in the body depends on the chemical form. In general, following oral ingestion, approximately 20% of radium is taken up by the intestines with the majority passing through the gut to be excreted in faeces and less than 5% excretion in urine. Once in the bloodstream, radium behaves in a similar manner to calcium and is rapidly taken up by the skeleton with relatively little uptake in the liver, kidneys or heart. The bone-seeking properties of radium resulted in the skeletal effects observed in those who drank the Radiothor health tonic. In addition to the carcinogenic effects on bone cells, the accumulation of radium in bone also irradiates the bone marrow which is a particularly radiosensitive organ at the centre of the long bones (arms and legs), the pelvis and spine. Irradiation of the bone marrow causes mutation and depletion of the blood-forming cells (red cells, white cells and platelets), resulting in a number of blood cancers such as leukaemia, lymphoma, myeloma and aplastic anaemia.

FIGURE 5.16   Radium girls working in a factory owned by the US Radium Corporation, circa 1922.

Source: Radium girls working in a factory owned by the US Radium Corporation, circa 1922. Photograph in the public domain

Furthermore, the suppression of white cells reduces the body's ability to fight infection, and the loss of platelets affects blood clotting resulting in an inability to stem bleeding.

Not long after starting work in the dial painting factories the girls noticed that their mouths would glow in the dark. A similar glow was seen from a handkerchief after nose blowing. Initially this was regarded as a harmless novelty and the girls even painted their teeth, fingernails, buttons and rings so they would sparkle at night. However, over time the absorption of radium led to terrible consequences. In 1921 some of the first health effects in the workers at the Orange dial painting studios were noticed. One of the workers, Amelia Mollie Maggia, was having painful dental problems and she had a difficult tooth extraction which never healed. Mollie had to quit her job in early 1922 when her illness worsened. The gums surrounding the tooth deteriorated requiring further visits to her dentist. During one visit, the bone disintegration was so bad that when her dentist gently used his fingers to examine her jawbone, it cracked and he was able lift it out from her mouth. Mollie died in September 1922. The death certificate stated that the cause of her death was ulcerative stomatitis (a rare condition of chronic mouth ulcers that is mainly found in late middle age white women) with complications from syphilis. Soon after similar effects were seen in dial painters at the Waterbury Clock Company (later known as Timex) in Connecticut and at the Radium Dial Company in Ottawa city, Illinois, which was run by Joseph Kelly and had 1,000 employees. In each case the death certificates stated causes such as anaemia, trench mouth, pneumonia and phosphorous poisoning. However, references to syphilis were part of a deliberate disinformation campaign to undermine the credibility of the workers and to divert any suspicions of deaths due to radium poisoning. Late in 1922 the New Jersey Department of Labour investigated the jaw necrosis cases at the Orange factory and after initially considering this to be due to phosphorus poisoning concluded that radium may have been the cause. In fact, the chronic intake of radium was eating away at the bones, causing bone tumours, crippling and painful skeletal deformities, painful tooth decay, gastrointestinal problems, chronic anaemia, loss of immunity and generalised body wasting.

Grace Fryer was another worker at the US Radium Corporation factory. She developed dental problems, a severely weakened jaw (later to become known as *Radium Jaw*) and spinal collapse which she was convinced had resulted from her work. She was the daughter of a union representative and had taken the job dial painting to do her part in supporting her brothers who had joined the army to fight in the war. Aware of the health problems she and her co-workers were suffering she sought medical opinion from Dr Frederick Flynn, someone who was recommended to her as a suitable medical specialist. After a thorough examination he and a colleague, who both claimed to be medical experts, stated that they found nothing wrong with Grace. It later emerged that Flynn was not a licensed physician, but a toxicologist, who secretly worked for US Radium Corporation. It took Grace a further 2 years to find a lawyer who would take up her case against the US Radium Corporation. In 1927 Raymond Berry and the Consumer's League of New Jersey took the case on behalf of Grace and four other dial painters, Edna Hussman, Katherine Schaub, Quinta McDonald and her sister Albina Larice, all who later became known as the Radium Girls. At the first court hearing on 11 January 1928, the women could not even raise their arms to take the oath due to their fragile condition. The court case attracted a huge amount of interest concerning legal precedents and employment rights, and after a 14-month delay by the Radium Corporation in 1928 they settled for an out-of-court award of $10,000 dollars plus a $600 annual pension and medical expense payments. The company never accepted liability and said that it settled not because the women were poisoned by radium but because it couldn't get a fair trial. Grace died on the 3 October 1933 at the age of 34. The legal challenge that she initiated ultimately led to the industrial regulations concerning occupational diseases and recognition that companies were responsible for the safety of their employees. Surprisingly though, the last radium dial studio in Ottawa, Illinois, which was called Luminous Processes and run by Joseph Kelly did not close until 1978. The final acknowledgement of the sufferings the dial painters endured prompted more detailed scientific study of human radiation effects and led to what eventually became known as the field of Health Physics. The increased understanding of radiation biology also fostered an improved safety culture among the workers on the atomic bomb in the 1940s.

5.8 RADIOTHERAPY

After the 1930s the inappropriate use of radium gradually ceased, making way for medical applications based on more rigorous scientific methods. Henri Becquerel had previously observed a severe skin inflammation under his waistcoat pocket where he had been carrying a tube of radium around with him. On showing this to a dermatologist the skin irritation was considered to be directly a result of the radium rays. In 1901 Becquerel obtained some more radium which he gave to Henri-Alexandre Danlos at the Hospital St Louis in Paris. Danlos successfully treated a few cases of lupus, an autoimmune condition where the body attacks its own tissues causing joint pain, skin rashes and tiredness. Early attempts by other researchers failed to reproduce these beneficial effects. For a number of years very few therapy trials were carried out since radium was expensive and difficult to get hold of. However, as greater amounts of radium became available it was seen as a promising treatment which could be applied in a variety of ways that X-rays could not. Early methods included radium emanation which used the radioactive gas, radon which is produced as radium decays to polonium. The gas was inhaled for the treatment of lung conditions and could be encapsulated in glass tubes which were used as applicators. Over time, as the science progressed, radium treatments were used to treat a variety of tumours and radium wards could be found in most hospitals. Radium applicators were used to give what has become to be known as brachytherapy treatments. Radium pads were attached to the skin for the treatment of skin tumours. Radium tubes were inserted into body cavities to treat cancer of the cervix and uterus and radium needles were used to treat tumours of the tongue and neck. As radium sources of higher activity became available, treatments could be delivered from a teletherapy machine known as radium bomb which contained about 5 grams of radium-226 (an early teletherapy unit is shown in Figure 5.17).

Compared to modern therapeutic use, this amount of radioactivity was comparatively low and the treatment procedures took a long time. Patients were given books

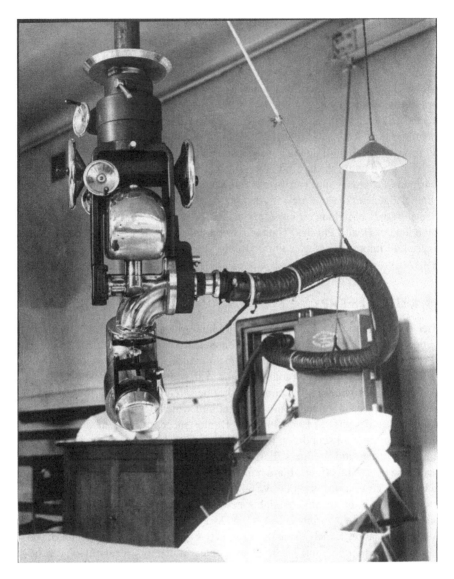

FIGURE 5.17   Radium bomb or teleradium unit containing 5 grams of radium used in 1947 for the treatment of head and neck cancer at the Fulham Road Cancer Hospital London. The lead safe containing the radium can be seen in the background.

Source: Radium bomb or teleradium unit. www.shutterstock.com/editorial/image-editorial/ medical-hospital-equipment-the-latest-type-of-radium-bomb-or-teleradium-unit-containing-5-grams-of-radium-used-in-the-treatment-of-cancer-of-the-head-ad-neck-at-the-cancer-hospital-in-fulham-road-london-the-safe-containing-the-radium-is-in-the-backgr-1736765a

and newspapers to read and cigarettes to smoke during the exposure (this was before doctors understood the harmful effects of smoking). Although these early procedures seem basic and lacking in directional ability to target diseased tissue, it should be appreciated that modern practice of radiotherapy that is used to treat so many patients today has developed from these early treatments. From the 1920s physicians began to understand that giving small, targeted radiation doses over time produced better results than a treatment with one large dose, since this allowed the normal tissues to recover between the treatment fractions. Between 1930 and 1950 radium and Supervoltage X-rays of energy up to 200 kV were used for treatment. To treat tumours at deeper depth in the body Megavotage X-rays and electrons were used. Radium sources were replaced with cobalt-60 and caesium-137 which produced higher energy gamma rays for teletherapy. In the 1950s the first high-power linear accelerators (often called a 'LINAC') were introduced for cancer therapy. The first medical treatment took place using an 8 MV LINAC at the Hammersmith Hospital in London in 1953. Today these devices use a tuned-cavity waveguide to accelerate electrons for either electron or X-ray beam treatments at energies up to 25 MV. Computer-controlled treatments are now given using personal image-guided plans and modulation of the beam energy to shape the treatment area for precision tumour therapy. The published statistics from Cancer Research UK show that 27% of cancer patients were treated with radiotherapy in 2020.

Given the past harmful biological effects of radium it might be thought that any future medical applications would be shunned, with radium treatments cast into the realms of history. However, in 2003 workers in the Department of Oncology at the Norwegian Radium Hospital in Oslo obtained a patent for the use of radium-223 to treat bone cancer and alleviate pain from metastatic bone disease (Larsen, Henriksen and Bruland. US Patent No.: US 6,635,234 B1). Artificially produced radium-223 with a physical half-life of 11.4 days decays by releasing a series of alpha particles. Injection of radium-223 in the form of radium chloride produced dramatic results in a series of controlled clinical trials. This bone-seeking formulation has now been licenced as Xofigo™ by the Bayer Pharmaceutics Company in Germany for therapeutic use in adult patients with metastatic castration-resistant prostate cancer and symptomatic bone metastases. This medicinal product is given as 6 injections each containing an activity of 55 kBq per kilogram body weight over a period of 6 months. This gives further support to the notion that a medicine is no more than a poison given at the correct dose.

6

Chain Reactions

Now I am become the destroyer of worlds.
Robert Oppenheimer, quoting the Hindu Bhagavad Gita scripture on reflection
of the heavy burden of creating the atomic bomb

6.1 EMBRACING THE ATOM

The discovery of natural radioactivity by Marie and Pierre Curie at the start of the 20th century opened a Pandora's box of scientific treasures. This was the start of a journey of further exploration into the nature of the atom by a succession of eminent physicists, many of whom earned Nobel Prizes for their efforts. In 1902 Ernest Rutherford showed that when radioactive atoms emitted alpha particles they changed to a different element. In 1919 he went on to direct alpha particles from a radium source at nitrogen, producing a rearrangement of the atomic nucleus to form oxygen. This was the first experiment to report a *nuclear reaction* which was written by the theoretical equation

$$^{14}N + a \rightarrow {}^{17}O + p$$

where a is an alpha particle containing two neutrons and two protons (p).

Rutherford's theory was confirmed in 1932 when experiments performed by the British physicist James Chadwick discovered the neutron, a particle having similar size to a proton but with neutral charge. Shortly after the neutron was discovered the Hungarian physicist Leo Szilard made a theoretical proposal that a nuclear chain reaction could occur as long as neutrons were released as the atoms divided. Also, in 1932 the British and Irish physicists John Cockcroft and Ernest Walton produced the first artificial nuclear disintegration in history using an electrically powered generator to accelerate protons towards atoms. Two years later in 1934 Marie Curies' daughter Irene and her husband Frederic Joliot found that these transformations could create artificial radionuclides. A year later the Italian physicist Enrico Fermi showed that a far greater variety of artificial radionuclides could be formed if neutrons were fired at atoms instead of protons. Whilst working in Berlin in 1938, Otto Hahn and Fritz Strassmann showed that uranium could be split into lighter elements, a process that was named 'atomic fission'. The picture became clearer when the structure of the atom was explained by the German physicist Niels Bohr, who modelled the atom in the form of a solar system with negatively charged electrons orbiting around the positively charged central nucleus. Working under Bohr, Lise Meitner and her nephew Otto Frisch then described the process of neutron capture as a severe vibration causing the nucleus to split into two not quite equal parts. They calculated that huge amounts of energy, of the order of 200 million electron volts, were released from the fission reaction. Frisch went on to carry out experiments in 1939 which confirmed these calculations. The final piece of information needed to create a nuclear fission chain reaction process, which could be used in an atomic bomb, was provided by the French theoretical physicist Francis Perrin when, in 1939, he introduced the concept that a critical mass of uranium was required to produce a self-sustaining release of energy. The theory and experimental basis of atomic science and nuclear fission was developed over a period of 45 years, between 1895 and 1939. Over the following 6 years driven by the necessities of war, American, British and Canadian collaboration devoted huge scientific and engineering resources to put this knowledge into practice.

On 1 September 1939, Adolf Hitler initiated the Nazi invasion of Poland. Two days later the UK and France declared war on Germany. Some years earlier, in the early 1930s when Adolf Hitler was appointed chancellor, the German government issued the *Gesetz zur Wiederherstellung des Berufsbeamtentums*, the Law for the Restoration of the Professional Service. This law ordered that those in government positions who had at least one Jewish grandparent or were political opponents of the Nazi Party should be dismissed immediately. Thousands of teachers, judges, police officers and academics at the country's top universities lost their jobs. Over the ensuing years hundreds of German scientists and intellectuals escaped to the UK and the US. This even included scientists like Frits Haber, who had become a German national hero by developing chemical weapons in the First World War. The list of escapees in the field of physics included Albert Einstein, Hans Bethe, Felix Bloch, Max Born, James Franck, Otto Frisch, Fritz London, Lise Meitner, Rudolf Peierls, Erwin Schrödinger, Otto Stern, Leo Szilard, Edward Teller, Victor Weisskopf and Eugene Wigner. Many of these physicists made significant contribution to the US atomic weapons project. The British were initially leading the way with the science, when in 1940 expatriate German physicists Otto Frisch and Rudolf Peierls, who were working at the University of Birmingham, showed that 10 kilograms of uranium-235 would be sufficient to form a critical mass capable of producing an atomic explosion. In 1941 James Chadwick produced a detailed technical report with the design and costs for the development of a bomb. In December 1941 after the Japanese bombed the American naval base at Pearl Harbour in Hawaii, the US entered the war. The possibility that Germany could build an atomic weapon caused great concern amongst the prominent scientists who were working in atomic physics. Two of these physicists Leo Szilard and Eugene Wigner, who were originally from Hungary, drafted a letter which warned of the potential development of a new type of extremely powerful bomb. They decided that Albert Einstein would be best placed to sign the letter due to his notoriety at the time. The Einstein–Szilard letter urged the US to take steps to acquire stockpiles of uranium and to accelerate the research of the Italian physicist Enrico Fermi and others. This was enough to convince President Franklin Roosevelt to authorise the work.

6.2 THE MANHATTAN PROJECT

The programme of work to design and build the atomic weapon originally had the code name of 'Development of Substitute Materials' but later took the name of the district of Manhattan, where the original offices used by the US army were sited, next to the Stone and Webster Engineering Company on Broadway. The groundwork for the Manhattan project had already been started in 1938 by a team of scientists working at Columbia University New York. The scientists, including Enrico Fremi, Leo Szilard, Walter Zinn and Herbert Anderson, were measuring neutron emissions during nuclear fission. Fermi, in particular, was described as a genius at creating both esoteric theories and elegant experiments. In 1938 he was awarded the Nobel Prize in Physics for the discovery of new elements from neutron bombardment, but he was forced to leave his Professorship in Theoretical Physics at the Sapienza University of Rome to escape from Benito Mussolini's new Italian government racial laws that affected his Jewish wife, Laura.

Early in 1942 Fermi and the team moved to Chicago and the first reactor was assembled in a disused squash court under the west stand of the Chicago Stagg Football Field. The reactor was known as Chicago Pile 1, since it consisted of uranium and uranium oxide in a cubic lattice of 40,000 graphite blocks piled together and held in a wooden frame (Figure 6.1). The graphite served as a moderating material to reduce the speed of neutrons, thus enabling a sustained chain reaction. The

FIGURE 6.1   Chicago Pile 1, the first nuclear reactor to achieve a self-sustaining chain reaction on 2 December 1942 built under the direction of Enrico Fermi.

Source: Chicago Pile 1. www.shutterstock.com/image-photo/first-nuclear-reactor-university-chicago-achieved-339962981

reaction process was controlled by a set of cadmium rods that were inserted and removed by hand. On 2 December 1942, following removal of the final control rod, the pile went critical and the nuclear reaction became self-sustaining over a few minutes until Fermi ordered the reactor shut off. This was the first controlled and sustained nuclear chain reaction, producing enough energy to power a light bulb.

Also, in 1942 Lieutenant General Leslie Groves of the US Army Corps of Engineers and son of a US army chaplain was appointed to take charge of the Manhattan Project. He was involved in most aspects of the development work and chose the sites for the research laboratories and production facilities. The most challenging aspect of the Manhattan Project was the manufacture and testing of the nuclear fuel, uranium-235 and plutonium-239. The site chosen for uranium-235 production and processing work was at Black Oak Ridge near Knoxville, later to be known as Oak Ridge (Figure 6.2). This had the advantage of being an isolated area in the Tennessee Valley with a plentiful supply of water for hydroelectric power and reactor cooling.

Once these research sites were chosen, federal agents, company managers and army personnel swooped in with the urgency accorded to a natural disaster. These remote villages and reservations were turned into atomic towns, as compulsory purchase of land and buildings was followed by removal of the residents, erection of fencing and

FIGURE 6.2   Aerial view of the K25 Plant at the Oak Ridge site of the Manhattan Project. This massive building housed the compressors and converters. 1947 photo by Ed Westcott.

Source: Aerial view of the K25 Plant. www.shutterstock.com/image-photo/aerial-view-k25-plant-oak-ridge-249574231

gatehouses to ensure security. The construction of plant, houses, shops and schools followed (Figure 6.3). Migrant and prison labour was used for constructing buildings and infrastructure but once the facilities were established key workers were brought in.

Uranium-235 production at Oak Ridge was undertaken by two methods, electromagnetic particle deflection and gaseous diffusion, but numerous setbacks and problems meant that only small amounts were produced. The Army and Navy also set up a third production process based on liquid thermal diffusion. Even with the three production streams, there was only enough uranium-235 for the construction of one bomb. Plutonium-239 was considered to be the most suitable material for use in the atomic bomb because it had a smaller critical mass than uranium and it could be produced in larger amounts in a nuclear reactor. Hanford near the Columbia River in Washington State was chosen as a further site for a reactor and processing facility to produce plutonium-239. This facility was managed by the DuPont Company, one of the largest US corporations at that time (Figure 6.4).

General Groves had to appoint an exceptional scientist with the interdisciplinary skills to manage the scientific and technological aspects of the work. When he met Robert Oppenheimer, a physicist from the University of California Berkley, he was impressed by his breadth of knowledge, not just in physics but also in chemistry, metallurgy and engineering. Oppenheimer was made director of the project to

FIGURE 6.3   Small homes under construction in Oak Ridge, Tennessee, for workers of the Manhattan Project's Clinton Engineering Works.

Source: Small homes under construction in Oak Ridge. www.shutterstock.com/image-photo/small-simple-homes-under-construction-oak-251929762

FIGURE 6.4   Aerial view of the 100-B Area of the Hanford Site, showing reactor B, the first large-scale nuclear reactor ever built. This produced plutonium-239 for the second atomic bomb which was dropped on Nagasaki.

Source: Aerial view of the 100-B Area of the Hanford Site. www.shutterstock.com/image-photo/aerial-view-100b-area-reactor-b-244388992

design and build the bomb despite some reservations from security officials as to his links with communist organisations. The project was so secret that the President of University of California, Robert Sproul, did not know the true nature of the research and was under the impression that the aim of the project was to construct a death ray device. The site that Oppenheimer and Groves selected for work on the bomb was at the Los Alamos Ranch School, 40 miles northwest of Santa Fe in the New Mexico desert. This location was remote, had a large area for outside explosive projects and a climate which allowed outside work all year round. The Los Alamos Ranch originally comprised 54 buildings, including 27 houses, dormitories and living quarters, 27 miscellaneous buildings, a public school, an arts and crafts building, a carpentry shop, a small sawmill, barns, garages, sheds and an icehouse. Once requisitioned, the area was designated as a military reservation and fenced off with security emplacements. The physics laboratory was built in an inner fenced and guarded compound known as the 'Technical Area'. By late 1944 a new building was constructed about a mile away from the Technical Area for all plutonium chemistry and metallurgy work. Known as D Building, this was one of the most elaborate and costly structures on the site. It was designed to minimise contamination of plutonium by dust particles in the air. According to official reports it had 5 miles of piping, a complex air-conditioning and air-washing system with electrostatic dust removal, as well as the essential laboratory services of water, air, gas and electricity. This was where the reactor produced polonium in gram quantities and was first fabricated into the pure metallic form needed to build the atomic bomb.

As the Manhattan Project progressed the site grew in size and a whole community developed, with thousands of people working and living at Los Alamos. The urgency of the war effort placed huge pressures on workers and the construction and engineering work in remote areas were not without risk. Carrying out scientific experiments in new and unknown areas of nuclear physics were even more hazardous. At Los Alamos there were 24 deaths between 1943 and 1946. The majority of accidental deaths occurred during the construction work, with recorded deaths including those from truck and tractor driving accidents, weapons malfunction, suicide, a fall from a horse and a child who drowned in a pond. Four unfortunate fatalities involved a group of janitors who shared muscatel wine that was laced with antifreeze. There were few early recorded deaths due to radiation exposure, since managers did not want to cause any concern to employees who worked with radioactive materials. The primary concern of General Groves and his team was to maintain the ongoing operation of the sites and not to allow anything to undermine secrecy. However, over the operational life of the project there was evidence of a number of incidents involving radiation. Accidents, human error, explosions and equipment failure such as broken seals, refrigerator failure and blocked pipes led to gas, particulate and liquid escapes. In some cases, these accidents caused near immediate death, but over time there was widespread environmental contamination of the land and water courses around the project sites.

6.3 THE PHILADELPHIA INCIDENT

In the autumn of 1944, an experimental laboratory process involving uranium hexafluoride was transferred to a production facility at the Naval Yard in Philadelphia. On September 2 a tube in the transfer room containing uranium hexafluoride and surrounded by concentric high-pressure steam pipes had become clogged. Three men who had volunteered to work in the facility went in to free the blockage. These were Peter Bragg a chemical engineer from Arkansas, Douglas Meigs an employee of the Ferguson Company, Cleveland, and Arnold Kramish, a physicist from the Special Engineer Detachment who was on loan from Oak Ridge. At 1:20 pm as Bragg and Meigs worked on their knees with a Bunsen burner to apply heat to free the clogged tube there was a sudden and violet explosion showering the men with steam, hydrofluoric acid and uranium hexafluoride. Both Bragg and Meigs were dead within minutes, their bodies covered in burns. Kramish was also critically injured with burns. In dock immediately alongside the transfer room the USS Wisconsin was rocked by the blast. The crew of the battleship and the firemen who responded to the explosion were never informed of the nature of the blast. Although the toxicity of uranium hexafluoride is more hazardous than the radioactivity, Groves drew a veil over the incident and the event was simply reported as an 'explosion at the Navy Yard'. It was not until many years later that the true nature of the incident emerged.

6.4 TICKLING THE DRAGON'S TAIL

The most dangerous scientific experiment of the entire Manhattan Project was to assess the critical mass. This was the amount of radioactivity needed to produce the nuclear bomb. It was a particularly risky operation, since there was no way of anticipating the dangers from any particular experiment until it was actually taking place. One key design aspect of the atomic bomb was to create an initial detonation that resulted in an implosion rather than a conventional explosion. This was to compress the fissile material to a critical mass which then resulted in an uncontrolled chain reaction. To determine the amount of material needed for the bomb the radioactive core was assembled by hand. The radioactivity was relatively safe to handle until it was in amounts sufficient for it to become supercritical. At this point the production of neutrons reached a threshold point where the chain reaction started to increase exponentially, reaching a state known as 'prompt critical'. As soon as the rapid increase in energy was measured, the core had to be separated to stop the critical reaction. One of the physicists on the design team was Harry Daghlian, who had previously studied at Massachusetts Institute of Technology and Perdue University. He was recruited to Los Alamos as a member of Otto Frisch's team, working on the critical mass of materials that would undergo fission to create a chain reaction. His contribution to these experiments had helped to prepare the plutonium core that was used at the first detonation which Oppenheimer called the 'Trinity test'. On 21 August 1945, Daghlian was attempting to build a neutron reflector around a spherical core of plutonium-239 using 4.4 kilogram tungsten carbide bricks. The purpose of the experiment was to determine if the neutron reflector would reduce the amount of plutonium required to achieve a critical reaction. After working during the day, Daghlian went to dinner and then in violation of the safety protocols decided to return to the lab to carry on alone with the experiment. The only other person in the area was Private Robert Hemmerly, a security guard who sat at his desk about 4 metres away. With a radiation monitor directed at the core Daghlian constructed four layers of bricks. This required a steady hand, as if he was building a 'Jenga block tower' with unstable nuclear bricks. After placing the fifth layer he was about to position another brick over the centre when his radiation monitor informed him that the core was about to go supercritical. While moving to remove the last brick he dropped it directly on top of the core, causing a supercritical chain reaction. In desperation to stop the reaction, he tried to knock the brick away with his hand, but with the knowledge that he had just received a lethal dose of neutrons he then had to dismantle the core block by block to render it safe. In doing so he suffered burns to his hand and further increased his exposure from the gamma radiation that was emitted (Figure 6.5).

This was the first criticality accident and the first severe radiation poisoning associated with the Manhattan Project. The Los Alamos medical team who were caring for Daghlian were not sure how to treat such poisoning and could only manage to alleviate his symptoms as best they could. His arm became severely swollen and blistered, and his blood cell count dropped. After a painful ordeal which lasted 25 days, Daghlian died on 15 September 1945. The medical record stated the cause of death was acute radiation syndrome with failure of the blood-forming bone marrow tissues. He was 24 years of age. It is difficult to calculate radiation doses

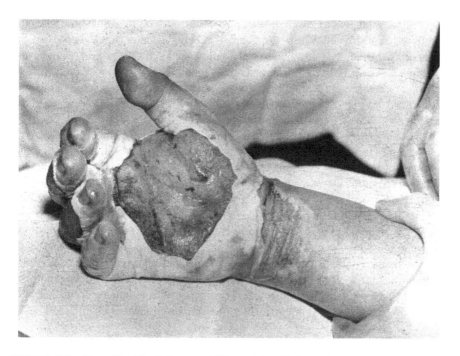

FIGURE 6.5   Harry Daghlian's burnt and blistered hand 9 days after he received his fatal radiation dose from the accident. Daghlian died from radiation poisoning 16 days after this photograph was taken.

Source: Harry Daghlian's burnt and blistered hand. Image in the public domain

following such accidents, but it has been estimated that Harry Daghlian received a dose of around 5 Sv, where the sievert is a unit of radiation absorption that takes into account the type of radiation and its biological effect on different tissues. Whilst mild symptoms are evident at doses over 0.3 Sv any radiation dose to the body over 0.5 Sv would result in an acute radiation syndrome, which basically causes nausea, vomiting and possibly diarrhoea. Half of the individuals exposed to 5 mSv would be expected to die within 30 days.

Private Robert Hemmerly, who was in the area at the time of the event, received a far lower radiation dose. He lived for another 33 years and died at the age of 62 of acute leukaemia.

A memorial stone to Harry Daghlian in Cullins Park, New London, Connecticut, reads

A brilliant scientist on the Manhattan Project. His work involved determination of critical mass. During an experiment gone awry he became the first American casualty of the atomic age. Though not in uniform he died in service to his country.

As a result of his death, stricter safety measures were implemented, but danger from such experiments remained. One of Daghlian's colleagues who also worked on the core assembly was a Canadian-born physicist, Louis Slotin. At the time when the bombs were constructed, Slotin was regarded as being the most expert person capable of handling the dangerous amounts of plutonium needed for the project. Both he and Daghlian had helped assemble the first atomic bomb that was dropped on Nagasaki. A photograph taken at the time shows Slotin standing beside the bomb looking cool and relaxed wearing a pair of sunglasses and with his shirt unbuttoned whilst those next to him are deep in concentration, as the device is assembled by hand (Figure 6.6).

On the afternoon of 21 May 1946, less than a year after the end of the war and 9 months after Harry Daghlian's unfortunate accident, Slotin, who was about to finish his time at Los Alamos, was conducting a criticality experiment to demonstrate the process to Alvin Graves, the physicist who would replace him in the bomb assembly team. He was working with the nuclear core which was originally constructed for use in a third bomb to be used on Japan but was never needed. At the end of the war it was returned to Los Alamos to be used in further experiments and post-war nuclear tests. Slotin was used to handling the dull metal neutron deflecting beryllium hemispheres and had done this procedure on numerous previous occasions. He lowered the half-shell of beryllium, called the tamper, over the core, stopping just before the two surfaces closed together. The tamper reflected the neutrons that were emitted by the plutonium, jump-starting the short-lived nuclear chain reaction. As

FIGURE 6.6 From the left, Herbert Lehr, Harry K Daghlian, Louis Slotin and an unknown member of the Los Alamos team, assembling the 'gadget' (the project name for the bomb).

Source: Los Alamos team. Image in the public domain

this was performed the physicists would gather the data in the same way they had when Daghlian was carrying out the critical experiment. Slotin held the reflector in his left hand and a long flat ended screwdriver in his right hand, which he used as a wedge to keep the two hemispheres apart. His technique was to rotate the end of the screwdriver to carefully lower the tamper to the point of criticality. This was a tricky operation known as *tickling the dragon's tail*, a dangerous manipulation that could wake something frightful if it went wrong.

At 3 pm, as Slotin slowly twisted the screwdriver lowering the tamper, it slipped on the metal surface causing the spheres to close together. Slotin was quick to react, twisting his wrist to flip the tamper away, but the damage was done and for a brief moment the 'dragon' woke. There was a sudden, intense blue flash as the air was electrified and a wave of heat passed through the room. Subsequent calculations have estimated that the chain reaction was a million times smaller than those of the bombs dropped over Japan, but this was still enough to send out a high burst of radiation. At the time there were seven other people in the room. One of them, Private Patrick Cleary, was a guard who was there to keep an eye on the plutonium and had little knowledge of what Slotin was doing. When he saw the flash and heard people shouting, he quickly ran out of the door and up a nearby hill. The other workers in the room included three physicists, a physics student, an engineer and a photographer. Most of them evacuated the room and an ambulance was called. Slotin was making his best efforts to determine where everyone was standing in the room and made a sketch to show their positions relative to the core. He attempted to use a radiation detector to measure the activity on various nearby items including a hammer, a tape measure, a brush and an empty Coca-Cola bottle. He failed to get any useful readings because the detector had been heavily contaminated and was affected by the electrical pulse from the radiation burst. He then instructed one of the physicists to place film badge radiation detectors around the area, further increasing his colleague's radiation exposure from the still active core. This also failed to produce any meaningful data on the nature of the incident.

All eight men were taken to the Los Alamos medical unit for a full health assessment. As would be expected, Slotin, who was working directly over the core, received the highest radiation dose and his was far higher than the dose Harry Daghlian took at the time of his accident. The high neutron dose that his body absorbed had actually reduced the radiation exposure to the people standing behind him. By the time Slotin had arrived at the medical unit he had already vomited. He underwent a medical examination by the doctors and then suffered several more bouts of sickness over the following hours. By the next morning he had stopped vomiting and his general health seemed to have improved, apart from his left hand which had been in contact with core which became increasingly painful. Over the next few days his hand developed large blisters with a waxy blue appearance. To help reduce the pain and swelling the doctors packed his hand in ice. His right hand, which had been holding the screwdriver, received less radiation and was not so severely affected. Slotin knew he was dying and made a call to his parents. The army flew them from Winnipeg, and they arrived in New Mexico 4 days after the accident. By the fifth day, Slotin's white blood cell count dropped, dramatically reducing his resistance to infection. His condition then declined rapidly. He was suffering abdominal pain and nausea, his heart rate and temperature fluctuated and by the seventh day he was in a state of confusion. His heart began to fail and his lips turned blue, so his doctors moved him into an oxygen tent. He lapsed into a coma and died 9 days after the accident, at the age of 35. The cause was recorded as acute radiation syndrome. His body was shipped to Winnipeg for burial in a sealed Army casket. It has been estimated that Louis Slotin received an effective radiation dose of 21 Sv. More detailed figures of the estimated radiation doses received by each of the participants in the experiment are given in Table 6.1.

This accident was 9 months to the day after Harry Daghlian's fatal accident with the same plutonium core. Until this point the core had affectionately been nicknamed 'Rufus'. After the two critical accidents it became known as the 'Demon Core'. Prior to the accident it was planned that the core would be sent to the Bikini Atoll, in the Marshall Islands, where it would be detonated in front of thousands of observers as part of the post-war nuclear tests called Operation Crossroads; however, in 1946 it was melted down and reformed into a new weapon. All further criticality experiments were put on hold until new safety procedures excluding hands-on experiments had been devised. One of the surviving physicists, Harry Schreiber,

---

TABLE 6.1

Details of the individuals present in the room at the time of the criticality accident on 21 May 1946. Data taken from the report written by Louis Hemplemann, Clarence Lushbaugh and George Voelz, submitted to the Conference on Radiation Accident Preparedness, Oak Ridge, October 1979

Name	Age	Role	Radiation dose	Outcome
Lois Slotin	35	Physicist	10 Gy neutron 1.14 Gy gamma	Died of acute radiation syndrome after 9 days
Alvin Graves	34	Physicist	1.66 Gy neutron 0.26 Gy gamma	Died of a heart attack after 19 years
Samuel Allan	26	Physics student and later a patent attorney	Not known	Refused to take part in any follow-up studies. Died 55 years later
Marion Cieslicki	23	Physicist	0.12 Gy neutron 0.04 Gy gamma	Died of acute myelitic leukaemia after 19 years
Raemer Schreiber	36	Physicist	0.09 Gy neutron 0.03 Gy gamma	Died of natural causes after 52 years
Dwight Young	54	Photographer	0.51 Gy neutron	Died of aplastic anaemia and bacterial endocarditis 29 years after the accident
Theodore Perlman	23	Engineer	0.07 Gy neutron 0.02 Gy gamma	Believed to have died 42 years after the accident. Cause of death not recorded
Patrick Cleary	21	Security guard	0.33 Gy neutron 0.09 Gy gamma	Killed in action 4 years after the accident while fighting in the Korean War

later designed a remote-control system with TV cameras to perform future experiments with the core. After this time all personnel were kept at a quarter of a mile away during criticality tests.

6.5 STIRRING UP TROUBLE

Cecil Kelly was an experienced chemical operator who had worked for over 11 years at the Los Alamos Laboratories from 1946 to 1958 with a break between 1949 and 1955. For about half of this time he worked on the processing of weapons-grade plutonium. On 30 December 1958 he was working in the plutonium processing facility at the Los Alamos site over a large tank where plutonium compounds were purified and concentrated by dissolving in a chemical solution. The liquid in the tank was normally at a concentration of 0.1 gram per litre, but for some reason on this day the concentration had been made 100 times higher. The total amount of plutonium was estimated to be around 3.3 kilograms in the upper layers of the solution, which was close to the critical level capable of creating a chain reaction. When Kelly turned on the stirrer in the tank a whirlpool formed in the liquid as the lower aqueous layer was drawn outwards to the walls of the vessel. The upper plutonium layer flowed into the centre of the vortex increasing the concentration to a critical amount. In a pulse that lasted no more than 200 microseconds a burst of neutrons and gamma radiation was released. At that moment Kelly was standing at the foot of a ladder, looking through a viewing window into the tank. The tank shook, moving sideways, and he was knocked to the floor. In a state of confusion, he turned the stirrer off and then on again and ran from the building. Two other operators had seen a bright flash of light followed by a soft bang. They ran to help and found Kelly in a confused and uncoordinated state shouting 'I'm burning up'. Thinking he had suffered a chemical accident they took him to a shower. Supervisors, radiation monitoring personnel and a senior nurse came quickly on the scene. Although Kelly was in shock and virtually unconscious, the nurse naively noted that he had a healthy pink complexion. In fact, it was later confirmed that what the nurse thought was a healthy glowing skin colour was due to erythema from radiation burns. As Kelly was taken to hospital the monitoring staff carried out an environmental survey using alpha radiation detectors to look for the presence of any plutonium contamination. Surprisingly this did not show much evidence of contamination. However, when they changed the monitors and surveyed for gamma radiation, they found very high readings and realised what had taken place, so quickly left the area. Kelly's initial collapse and loss of mental capacity is now understood to be the first clinical stage of acute radiation syndrome. By the time he arrived at the Los Alamos Medical Centre he had dropped into semi-consciousness and was hyperventilating, retching and vomiting. His skin had a reddish violet colour and his lips were blue, this being a typical clinical sign of poor blood oxygenation. About 10 minutes after arrival Kelly's blood pressure was 80/40 and his heart rate was 160 beats per minute. He was shivering with uncontrollable movement of his limbs and torso. Radiation monitoring in the medical bay showed that his body was emitting gamma rays and radioactivity was detected in his vomit and faeces. The neutrons emitted during the critical event had activated some of the elements his body, such as sodium which had turned to sodium-24. Blood samples taken after

6 hours showed depletion of white cells. After 24 hours a bone marrow sample was taken from his sternum (breast bone), which showed that the marrow was watery with swollen cells, lacking normal appearance or function. Over the following day the pain in Kelly's stomach became more severe and he became increasingly restless. His pulse became irregular and he became pale as his perspiration increased. Kelly died around 35 hours after the incident. His radiation dose was estimated to be 36 mSv. His post-mortem examination was authorised by the Los Alamos coroner who was also the pathologist at the Los Alamos Medical Centre, Dr Clarence Lushbaugh. During the post-mortem examination tissue samples were taken from throughout Kelley's body to measure plutonium concentrations. The exposure to neutrons and gamma rays during the accident had no impact on the amount or distribution of the plutonium in his body. The tissue analysis showed that Kelley's total plutonium body burden was 650 Bq with about 50% of the plutonium in the liver, 36% in the skeleton, 10% in the lungs and 3% in the respiratory lymph nodes. This was subsequently considered to be from chronic intake of plutonium during his time working at Los Alamos.

After the Kelley incident, Lushbaugh decided to make the collection of tissue specimens for plutonium analysis a routine part of all post-mortem procedures performed at the Los Alamos Medical Centre. That practice was considered acceptable since in those days, post-mortem examinations were considered to be a learning tool to confirm the accuracy of the doctor's diagnosis and the effectiveness of certain medical treatments whilst also determining the cause of death. At that time post-mortem examinations were undertaken on a set percentage of patients who died in the hospital each year to maintain the accreditation of the hospital and the medical staff.

Lushbaugh subsequently attempted to get permission to perform a post-mortem examination on every person who died at the Los Alamos Medical Centre, including laboratory workers and members of the general population from the Los Alamos region. In most cases relatives agreed to the procedure but they were never aware that tissues were being retrieved for the assay of plutonium content. Procedures for obtaining formal consent of employees who were occupationally exposed to radiation were put in place in 1968 when the Atomic Energy Commission established the National Plutonium Registry.

6.6 DAWN OF THE ATOMIC AGE

The first indication that the work of the Manhattan Project was a success was when the first detonation of a nuclear device took place during the 'Trinity Test' which was carried out at the Northern end of the Alamogordo Bombing Range in the New Mexico desert at 5:29 am on 16 July 1945. The huge scientific and technological effort of the Manhattan project had involved over 130,00 people and a cost of $2 billion (equivalent to over $23 billion today). The uranium-235 from the Oak Ridge facility was used in the construction of the first bomb 'Little Boy' which was dropped on Hiroshima on 6 August 1945. Three days later on the August 9 the second bomb 'Fat Man' containing plutonium-239 from the Hanford site was dropped over Nagasaki. It is estimated that the nuclear explosions and radiation killed between 129,000 and

226,000 people. Thankfully, to this day these two bombings remain the only nuclear weapons to be used in armed conflict. These acts by the US military resulted in the Japanese surrender which brought an end to the Second World War. Many people argue that with no end in sight to the war in the Pacific, this prevented the loss of many lives of the American and allied forces and civilian populations in the Far East.

The immediate biological hazards from nuclear explosions are a bright flash which can cause temporary blindness, followed by an intense blast wave capable of causing death, injury and extensive damage to buildings. Fire and heat following the blast wave results in severe burn injuries and the further destruction of buildings. A nuclear explosion also produces an electromagnetic pulse that can damage and disrupt power sources and electronic equipment over an area of several square kilometres. In addition, nuclear bombs produce large amounts of radiation, which are of two types. The critical event produces an intense release of neutrons, alpha particles, beta particles and gamma rays, that can cause whole body irradiation with subsequent damage to body tissues. Radiation exposure of this kind can cause the rapid onset of radiation sickness. The second form of radiation occurs after the explosion as the clouds disperse and the radioactive materials propelled into the upper atmosphere fall back to the ground. The Japanese survivors described the fallout pouring down to the earth, as 'black rain'. This contained radioactive materials such as cesium-137, iodine-131, americium-241 and strontium-90. These radioactive elements were inhaled, became ingested and caused the widespread contamination of people and the ground area. The radioactivity entered water supplies and was taken up by plants and livestock producing long-term issues that required remediation and management to reduce the levels of contamination.

The development of atomic weapons had changed the course of history. When news of the first successful detonation over Hiroshima reached the team at Los Alamos, the scientists were elated. When they learned of the huge loss of life and the scale of devastation this feeling was followed by the conflicting emotions of achievement and guilt. Nuclear physics had become a powerful force for governments in both the East and West. After the war there was intense debate between the US government, the military and scientific communities as to whether to proceed with development of the far more powerful weapon known as a hydrogen super bomb. Robert Oppenheimer who became a government advisor was against continuing work on the super bomb, as he considered this to be unnecessary and had the potential for huge loss of life. In 1954 following an FBI investigation, he was made to attend a security hearing concerning his associations with the communist party and suspicion that he was a Soviet spy. The outcome of the hearing was that Oppenheimer's security clearance was revoked and he was publicly humiliated, although he continued with his academic work, teaching and publishing. In his talks and writing he drew attention to the difficulty of managing the power of knowledge in a world in which academic freedom and the exchange of scientific ideas was increasingly constrained by political concerns. Oppenheimer was a chain smoker and died of throat cancer in 1967 at the age of 62.

After the war the US hoped to maintain a monopoly on the new weapon and although the UK had given them the original designs for the bomb, they did not share any details of this new technology. Nevertheless, the secret information for building the atomic bomb could not be contained. Four years after the bombs were dropped on Japan, the Soviet Union conducted its first nuclear test explosion. The UK, France and China followed. Those countries with nuclear weapons 'know how' continued to test weapons as part of the nuclear arms race. Global levels of atmospheric radiation reached a peak in the 1960s, but although the amounts have continued to decline since then it is estimated that there are still several tons of plutonium circulating in our biosphere. Seeking to prevent the expansion of nuclear weaponry the US and other like-minded countries negotiated the Nuclear Non-proliferation Treaty in 1968. Concerns regarding atmospheric radiation resulting from these tests led to the United Nations International Test Ban Treaty of 1996; however, there are at least eight nations that have failed to sign up to this and some countries such as North Korea have continued to carry out underground nuclear weapons testing. Currently it is estimated that throughout the world there are over 13,000 nuclear warheads, 90% of which are under the control of Russia and the US.

7 ☢

Nutopia

Medical and scientific progress depends upon learning about people's responses to new medicines, to new cutting-edge treatments, . . . but there is a right way and a wrong way to do it.
President Clinton apologising on 4 October 1995 to the survivors and families of those who unknowingly were subjects of government-sponsored radiation experiments.

7.1 HUMAN RADIATION EXPERIMENTS

The advances in science and technology brought on by the war between 1939 and 1945 produced much that improved life in the years that followed; however, both during and after the war there were instances where science was used for more evil purposes than good. The Second World War was the deadliest war in history, with an estimated loss of life of between 15 and 20 million in Europe alone. Over three million prisoners died in the Nazi prison camps and those that escaped death only did so if they were considered fit enough to work. As part of the programme of 'ethnic purification' the Nazis considered it necessary for these workers to be rendered incapable of propagation and to further these aims unethical human experiments were carried out by the SS physicians into methods that could be used for mass sterilisation. The SS leader, Heinrich Himmler, funded two competing experimental research programs: one looking at various poisons and the other using radiation for sterilisation. Himmler provided staff, funds and equipment and authorised experiments to be carried out on unwilling victims from the prison camps.

Two notable physicians working in the death camps were Professor Dr Carl Clauberg and Dr Horst Schumann. Clauberg developed a method of non-surgical mass sterilisation by introducing a specially prepared chemical irritant into the female reproductive organs. This produced severe inflammation of the fallopian tubes finally blocking them resulting in sterility. These experiments caused the death of some of his subjects whilst others were deliberately killed so that the effects could be examined at post-mortem.

Dr Schumann did not have any particular qualification for medical research. His past experience involved overseeing the Nazi killing centres and the selection of victims. Schumann was interested in 'scientific methods' for mass sterilisation and set up X-ray apparatus in Block 30 of the Birkenau Concentration Camp to carry out his experiments. Viktor Brack, the head of Hitler's Chancellery, was a supporter of Nazi research into sterilisation by radiation since he felt that surgical methods were too long and expensive. He reported to Himmler to whom he sent a letter in 1941 suggesting that the victims could be summonsed to attend special centres to complete obligatory but sham regulatory forms. The individuals would be made to sit at a counter in front of concealed X-ray apparatus whilst completing the documents. In this way they would be unaware that any procedure was being carried out. By 1942, the doctor and his assistants were at work on X-ray sterilisation experiments. It is estimated that around 1,000 Jewish prisoners were brought in for experiments to determine the amount of radiation required to cause sterility. The X-ray beam was directed at the women's ovaries and the men's testes for various time periods and at varying voltage and current settings to vary the dose. Higher exposures resulted in burns to the lower abdomen, groin and buttocks and festering sores developed which became infected. After the irradiation the unfortunate subjects underwent surgery to remove the ovaries and testes for microscopic evaluation. The men were also subject to other invasive procedures to extract semen. Many of the victims died from complications following the surgical procedure and the survivors of the experiments in their weakened condition were less likely to survive any subsequent enforced work details. In the end the results of the X-ray sterilisation procedures were not as the researchers expected. Schumann sent a final written report entitled 'The effect of X-ray radiation on the human reproductive glands' to Himmler in April 1944. His conclusion was that surgical castration was quicker and more reliable than X-ray irradiation.

7.2 TESTING TIMES

Experiments involving radiation and radioactivity continued in the post-war years as part of the cold war efforts of both the East and West. It has been estimated that since 1954 around 29,000 Hiroshima-sized bombs have been detonated either on the ground, at the top of towers, in underground mine shafts, on barges, underwater, dropped by aircraft, suspended from balloons and by firing in rockets into the atmosphere. These tests have caused one of the major sources of exposure from artificial radioactivity to the human population.

In Soviet Russia, Joseph Stalin originally had little enthusiasm for research in atomic physics and had instructed his scientists to concentrate on mining and metallurgy and to deal with army technical problems. In 1942 Russian physicist Georgy Flyorov noted that Western physicists had stopped publishing papers in atomic physics whilst publications in other areas of physics flourished. He wrote to Stalin suggesting that atomic physics had become a secret area of research and therefore needed more attention. Stalin promptly reallocated physicists to carry out research into nuclear reactions, which then became a Soviet Army project. But after a great deal of initial effort Soviet work on a uranium bomb made little progress. After the atomic bombings on Japan in 1945 the Soviets continued their atomic weapon research with a renewed sense of urgency. Progress was finally made with the development of a plutonium bomb based on British information obtained by NKVD secret police spy security agents.

In 1949 Lavrentiy Beria, the deputy prime minister under Stalin and the political head of the Soviet atomic bomb project, selected a large 18,000 square kilometre area of steppe in northeast Kazakhstan for the site for nuclear weapons testing. This area later became known as 'The Polygon'. The science buildings were located around 150 km west of the town of Semipalatinsk now known as Semey. Gulag camp labour was used for construction work. From 1949 until 1989 the Soviet Union conducted 456 nuclear tests at the Semipalatinsk test site, with little regard for their effect on the local people or environment. Beria had a disregard for the well-being of anyone living in the area claiming that it was uninhabited. The full impact of radiation exposure on the local inhabitants was hidden for many years by Soviet authorities. In 1991 after the fall of the Soviet Union the test site closed and fell into neglect. This abandoned nuclear facility raised major international concerns, since part of the area contained many underground test sites and a network of tunnels and boreholes which contained large amounts of fissile materials and weapons-grade plutonium. It was soon realised that these virtually unguarded sites were high on the list of the most radioactive places on earth and they were open to scavengers, terrorists and rogue states. There was widespread international consensus that this site had become the largest security risk since the fall of the Soviet Union. It took a 17-year combined effort by scientists and engineers from Kazakhstan, Russia and America to secure the site, at a cost of around $150 million. This work involved removing any intact stable radioactive materials and sealing tunnel entrances by pouring concrete mixes with the capacity to bind the plutonium waste and was completed in 2012. The Semipalatinsk area has become the most intensively researched atomic testing site in the world. Scientific studies have included the analysis of blood samples taken from over 40 families living in fallout areas. These identified increased levels of DNA mutations compared to those in a control group. Despite extensive scientific studies it has been difficult to make firm conclusions on radiation-induced health effects. One longitudinal study conducted over 40 years found a correlation between radiation fallout exposure and the prevalence of solid tumours of the oesophagus, stomach, lungs, breasts and liver; however, these are still subject to further investigation and discussion.

In September 1954 the Soviet Army carried out a military exercise code named 'Snowball' to study defensive and offensive operations during a nuclear war. This involved detonating a 40-kiloton nuclear bomb, twice the size of that of the bomb dropped on Hiroshima, at a height of 350 metres over a test area to the north of the village of Totskoye, which was west of the Semipalatinsk test area. The residents of the nearby villages were evacuated and given property insurance and a daily reimbursement for the duration of the exercise. Those who decided not to return to their homes were given new houses or financial compensation. Any residents who refused to leave their homes were told to dig ditches to reduce the effects of the blast. Following the explosion, a reconnaissance plane was sent to determine the movement of the radioactive cloud and map out the areas of highest contamination. An army of 45,000 soldiers with planes, tanks, armoured personnel carriers, cars, guns and mortars were then deployed to the area. The participants were selected from Soviet servicemen and were given 3 months' salary if they agreed to take part in the exercise and swear an oath of secrecy about the exercises with this new weapon. The soldiers were given suits, respirators, gas masks, gloves and tinted goggles. Individual dose meters were used to assess the levels of radiation received by the personnel. The soldiers were kept about 500 metres away from the hypocentre to avoid the most dangerous area of radiation. According to the data from remote control and measuring equipment installed at 730 metres from the hypocenter, the level of radiation reached 650 mSv per hour 2 minutes after the blast, dropping to 15 mSv per hour after around 50 minutes. In most regions of the world people would normally be exposed to between 1.5 and 3.5 mSv per year. Soviet scientists received detailed reports after the exercise which provided information on the impact of the nuclear blast on houses, shelters, vehicles, vegetation and experimental animals affected by the explosion. Unfortunately, the long-term health effects of those involved were not adequately assessed. Some participants claim they were not given protective clothing and that the high seasonal temperatures of over 40°C in the area made it impossible to wear gas masks. It was also claimed that after the manoeuvres there was insufficient attention to the safe removal and disposal of the contaminated clothing.

Another site for nuclear weapons testing was later chosen on the islands of Novaya Zemla in the Arctic Ocean in northern Russia. Over 220 nuclear tests have been carried out on these islands since 1954, including the largest nuclear weapon ever tested by the Russians. The last detonation on the islands was in 1990, but the Russia Ministry for Atomic Energy has carried out a series of subcritical underwater explosions since then. The islands and its airfields remain in military use today.

It should be appreciated that the Soviets were not alone in exploiting military personnel for radiation experiments. From 1946 to 1963, the US military ordered more than 200,000 soldiers to observe one or more nuclear bomb tests, either in the Pacific or at the Nevada Test Site. Most of the American test explosions took place during the night or before the sun rise at dawn. The first thing that followed the detonation was the intense light flash, which caused temporary blindless as the soldiers watched (Figures 7.1 and 7.2). Many observers wore dark goggles or turned away. The witnesses recalled that the light was so bright they could see the bones, muscles and veins beneath their skin. Once the light flash faded the fireball appeared and the sound of the blast could be heard, followed by a shockwave of overpressure which knocked many people over. Many of these servicemen never told anyone what they witnessed, since they had been sworn to secrecy. They were warned that they could be fined up to $10,000 if they told anyone that they were involved in the nuclear tests. Some soldiers feared that if they spoke about what they had seen they could be tried for treason and shot. It was only many decades later that they shared their experiences. Although there were many anecdotal reports of adverse health effects suffered by the military observers, it has been difficult to determine accurate and meaningful information concerning the extent of radiation exposure and subsequent medical outcomes.

As part of their weapons programme US forces detonated a total of 23 weapons on the Bikini Atoll in the Marshall Islands. In some tests the size of the bomb was miscalculated and the explosion was much larger than intended. This resulted in the destruction of the measuring instruments and more widespread radioactive contamination from cloud condensation and fallout. The Castle Bravo Test of a dry fuel thermonuclear bomb on 1 March 1954 was 1,000 times more powerful than the bombs that were dropped on Hiroshima and Nagasaki and was the largest US bomb ever tested. When this was detonated the US forces were not expecting such a

FIGURE 7.1   Men of the US Army, 11th Airborne Division, watch a plume of radioactive smoke rise over Yucca Flats during Exercise Desert Rock on 1 November 1951.

Source: Men of the US Army 11th Airborne Division. www.shutterstock.com/editorial/image-editorial/historical-collection-10292905a

large explosion or such widespread nuclear fallout that spread across the islands of Rongelap and Rongerik. Some of the island's inhabitants who developed symptoms of acute radiation syndrome were not evacuated until 3 days after the explosion.

Between 1960 and 1996 the French government carried out more than 200 nuclear weapon tests. Whilst they always claimed the operations were carried out as safely as possible, some defence researchers have stated that foot soldiers stationed in Algeria in 1961 were intentionally exposed during the explosions to gain knowledge of the human effects of the weapons. Similar reports were made in the case of Australian servicemen who were present during the British nuclear weapons tests at Emu Field, Maralinga, Montebello and Christmas Islands in Australia between 1952 and 1963. It is also understood that over 100,000 Chinese troops were sent into the deserts of Xinjiang Uyghur Autonomous Region to provide the labour for China's first atomic bomb test code name 'Project 596' at Lop Nur in 1964. It was reported that a number of these troops later developed serious medical problems.

FIGURE 7.2   Canadian and British observers illumined by the sun and the flash
from the atomic blast of ABLE, the first air-dropped nuclear device to be exploded on
American soil. The 1 kiloton bomb was dropped by a B-50 Superfortress aircraft on 27
January 1951 at Frenchman Flat, a dry lakebed in the Nevada Desert, 65 miles northwest
of Las Vegas. Most observers had goggles, but others without eye protection stood with
their backs to the blast.

Source: Canadian and British observers. www.shutterstock.com/editorial/image-editorial/historical-
collection-10292890a

7.3 ULTIUM, EXTREMIUM AND PLUTONIUM

The Manhattan Project produced a number of artificial radioelements which had
new and promising possibilities, but the one that received the most attention was
plutonium, which was the fissile material of choice for the atomic bomb. In 1940
Glenn Seaborg and Edwin McMillan used a large particle accelerator called a cyclotron, at the University of California Lawrence Radiation Laboratory in Berkeley, to
fire deuterons (the nucleus of heavy hydrogen containing one proton and one neutron) at uranium-238 to produce neptunium-238. Neptunium-238 then decayed to
plutonium-238 which has a half-life of 87 years and produces large amounts of heat
as it decays. This heat source proved to be suitable for electric power generation,
and since plutonium-238 emits alpha particles and only low levels of gamma rays
there was minimal need for heavy shielding around devices. NASA, for example,
has used plutonium-238 source for thermoelectric power generation in space craft.
The Mars Viking 1 and Viking 2 missions were powered by plutonium-238 thermal

generators and the Curiosity Rover Science Laboratory that landed on Mars in 2012 contained a 4 kilogram plutonium-238 power source. One of the most ambitious space missions attempted was the 20-year long Cassini-Huygens research mission to Saturn. Launched in 1997, the probe contained three plutonium-238 thermal generators that provided sufficient power for a 7-year flight to Saturn and 13 years in orbit. The Perserverance Rover that landed on Mars in 2021 is powered by 4.8 kilograms of plutonium dioxide; however, with production sites now closing, shortages of plutonium-238 have been raised by NASA as a major problem for future space missions.

In 1941 Glenn Seaborg, Joseph Kennedy, Arthur Wahl and Emilio Segrè went on to demonstrate that neptunium-239 could be produced by firing neutrons at uranium-238 which rapidly decays to plutonium-239 with a half-life of 24,100 years. Plutonium-239 proved to be far more suitable for use in explosive chain reactions. Seaborg and his team at Berkeley originally considered the names 'ultimum' and 'extremium' for the new metal they had discovered, since they originally thought this to be the at the very end of the periodic table. In line with naming other elements after the planets the name 'plutium' was considered, but they finally settled on the name 'plutonium' after the planet Pluto. Pluto is also the ruler of the underworld in Greek mythology.

Plutonium is a soft, silvery metal that is highly reactive. This high chemical reactivity is one of the reasons why plutonium is difficult to work with. It oxidises so easily that a fine powder of the pure metal will burn in air spontaneously. To prevent this, solid metal plutonium is coated with a thin layer of an alloy, such as platinum–rhodium, which stops oxygen in the air from reacting with the metal. This allows it to be handled and avoids the shedding of any radioactive particles of the metal that would contaminate anything that came into contact with it. The coating also absorbs most of the alpha radiation allowing it to be handled in small amounts. The harmful health effects of plutonium are due to both the radiation and the potential for heavy metal poisoning. The biological action of plutonium has been determined from information obtained following accidents, fires, detonations and other human experimentation. As plutonium decays alpha, beta and gamma radiation are produced, all of which can cause serious harm following both acute and long-term exposure. As with all forms of radiation, greater levels of exposure cause more serious health effects, including radiation sickness, genetic damage and cancer. Whilst gamma rays can penetrate into the body far greater harm is caused by ingested or inhaled plutonium. The most harmful effects have been shown to be due to inhaled plutonium depositing in the lungs, where it can cause lung cancer. If swallowed it is poorly absorbed into the body with only 0.04% of plutonium oxide taken up by the digestive tract. Absorption through the intestines and cell membranes occurs very slowly and once in the bloodstream it will gradually concentrate in the bones and liver. Metabolism and excretion are also slow with a biological half-life of 200 years.

The scientists at Los Alamos knew that working with plutonium would be very hazardous. Starting with small milligram amounts, they had to handle increasing larger quantities to make the metallic fuel for the bomb. As they started work nobody had ever seen pure plutonium metal or knew any of its physical or chemical properties. Nobody knew its density, its melting point, or how hard or brittle it was. Nobody knew how to fabricate it. All they knew was that they had to do it and as carefully as they could.

7.4 AN UNEXPECTED BITTER TASTE

There were a number of incidents where individuals were exposed to plutonium in either liquid or gaseous form; however, the first person to get real taste of plutonium was Donald Mastick, a 23-year-old chemist who was handling some of the first samples to arrive at the Los Alamos laboratory. On 1 August 1944 Mastic was given a small sealed glass vial containing 10 milligrams of plutonium in the form of a plutonium chloride solution. No one on site was aware that during transport and storage a chemical reaction, possibly enhanced by the alpha radiation in the solution, was causing internal pressure to build up in the glass vial. Whilst Mastick was standing at the workbench the top of the vial suddenly burst and a jet of liquid plutonium ejected from the opening and splattered on the wall in front of him, only to be sprayed back in his face and into his mouth. Not only was this a setback for the war effort, but Mastic realised that he had just swallowed some of the material to be used for making the atomic bomb. This was an unexpected shock that had left him with an acid taste in his mouth. He went straight to see the Los Alamos health director Louis Hemplemann, who promptly gave him a mouth wash and pumped out Mastick's stomach to retrieve as much of the swallowed material as he could. His face was thoroughly scrubbed, but the skin remained contaminated with about a microgram of plutonium. Despite the mouth wash and stomach pumping for days afterwards he could blow at an ionisation chamber and cause the needle to go off-scale. The level of internal contamination was later estimated to be about 10 micrograms, but since absorption through his gut would have been low his retained body burden was thought to be of the order of 1 microgram. After the wash up Mastic was considered fit to work and being a chemist, he was given the task of recovering the plutonium from the various washing solutions and the stomach contents to be put back into use for the bomb project. He later worked on the assay of plutonium in urine samples and subsequently transferred to work on assembly and testing the bombs dropped on Japan. Around 50 years later when Mastic was followed up at interview in 1995, he reported that he had experienced no ill effects from the incident. There is no reason to doubt his word on this. It may be that the prompt action to pump his stomach directly after his accident prevented most of the plutonium that he swallowed from being taken up in his bloodstream and reducing his long-term radiation dose.

7.5 THE US PLUTONIUM STUDIES

When the Manhattan Project took over the atomic weapon programs, it set up its own Medical Office. Stafford Warren from the University of Rochester, New York, was appointed Director and he established medical, health physics and biological research sections at Rochester, Los Alamos, Oak Ridge, Chicago and Berkeley California. The heads of the sections were Stafford Warren for the Manhattan Project and Oak Ridge, Louis Hemplemann at Los Alamos, Robert Stone at the Chicago Metabolic laboratory and Joseph Hamilton for the animal and human studies at Berkeley California. All four of these men were medical doctors with strong backgrounds in radiology. The medical division of the Manhattan Project grew to over 70 officers who monitored and cared for tens of thousands of staff. This work also included health physics research on individuals and environmental monitoring of the environment surrounding the project sites, including the air, rivers, lakes, livestock and agricultural produce.

Following Mastick's accident in 1944, Louis Hemplemann became increasingly concerned about worker safety. He was aware that they had no means of detecting how much plutonium had been retained in Mastic's body. He then requested permission from Robert Oppenheimer to develop methods for measuring plutonium in the lungs and biological samples including urine and faeces, so that the amount of radioactivity in the body could be assessed. Oppenheimer agreed that this was an important area of research and that methods for measuring the radioactivity in biological samples were necessary. In August 1944 Hemplemann met with Stafford Warren in his capacity as the Chief Medical Director of the Manhattan Project and they discussed Mastick's accident. They quickly realised their lack of understanding of the potential human biological effects of plutonium uptake in the human body and the military implications of what had happened. The increasing importance of the bomb project and the future possibility of the military deployment of atomic weapons raised the possibility of the widespread dispersal of radioactive materials. Although the harmful effects of X-rays and radium were known, senior officials were worried that the health effects of external irradiation, contamination and internal ingestion of these radioisotopes were unknown. Also, in 1944 Glenn Seaborg wrote to Robert Stone expressing concern about the health of his staff who were handling increasingly large amounts of plutonium. He requested that material should be made available to allow Joseph Hamilton to carry out distribution studies in the Chicago Met Labs. The medical team now had serious concerns as to how poisonous plutonium was and were concerned that those handing the material may be ingesting dust particles through the nose or mouth and possibly by absorption through the skin. It soon became apparent that the military and government should understand the toxic effects of this new metal and how radioactivity and radiation affected the human body. Stafford Warren concluded that scientific experiments were necessary to gain an understanding of these effects and a programme of research in both animals and humans was needed. Louis Hemplemann was asked to outline a work programme, which was to be carried out in three parts. The first task was to develop methods to determine the amount of plutonium excreted from the body in urine and faeces and establish techniques to measure radioactivity in body tissues and especially the lungs. Secondly animal experiments were to be carried out to validate the analytical methods that were developed. Finally, human studies would be carried out to measure the rate of plutonium excretion from the body.

Oppenheimer was concerned that the Los Alamos site had taken on too much work and was already far too overcrowded and not equipped for biological studies. In the spring of 1944, plutonium was made available for Robert Stone to undertake animal studies at the Chicago Met Lab where research was initiated on determining the acute toxicity of plutonium. These studies involved the injection of microgram and milligram quantities of plutonium-239 into mice, rats, rabbits and dogs. Samples of plutonium were also made available to Joseph Hamilton in Berkeley. Hamilton was both a Professor of Medicine and Radiology at University of California, San Francisco, and one of the first to become a Professor of Medical Physics at University of California Berkeley. He undertook experiments to assess the uptake of plutonium dust from the lungs of laboratory rats and soon found that about 25% of the inhaled plutonium remained in the lungs and around 20% eventually became deposited in the skeleton. The uptake of plutonium in bone was similar to that seen years earlier in those people who had ingested radium. These observations were not surprising since these elements are both heavy metals. The rest of the radioactivity was slowly eliminated from the body. The final amounts deposited and cleared depended on whether the plutonium compound was soluble plutonium nitrate, which was quite readily absorbed, or plutonium oxide which was poorly absorbed. By the end of 1945 studies with rodents and dogs had shown that the acute radiation effects of plutonium were less damaging than highly toxic chemicals, such as curare, strychnine and botulinus toxin, but they far exceeded any known chemical hazards of other heavy metals. The clinical observations of acute plutonium toxicity in dogs was quite similar to the effect of a single whole body lethal dose of X-rays. The main difference was that the initial vomiting and reduction in blood cell counts seen with X-ray irradiation were absent. Following plutonium administration refusal of food and water and weight loss in the first days were followed by a shock phase around day 10, which included a rise in body temperature and pulse rate, laboured breathing and internal bleeding. Changes in the blood were seen including the depletion of white and red cells. The experimental work only extended over a period of 6 or 7 months, so any medical opinion on chronic plutonium toxicity was essentially guesswork. A few experiments carried out in laboratory mice and rats had shown the development of bone tumours. The medical team set an initial figure of 30 micrograms as a potentially lethal intake of plutonium, with a tolerance limit of 5 micrograms.

By January 1945 Wright Langham at the Los Alamos Science Lab had developed a chemical method for detecting one ten-thousandth of a microgram of plutonium in urine and faecal samples. The first experimental measurements were carried out on the laboratory staff who were handling it on a daily basis. This was carried by taking nose wipes and measuring the levels of plutonium excreted in urine samples. The early results of the laboratory workers urine analysis showed levels much higher than expected. After further refinement of the collection techniques it was found that the urine samples were contaminated by particles from the workers clothing and skin. This meant that high environmental contamination from plutonium dust in the laboratory areas was much more widespread than originally appreciated. Plutonium was not just restricted to contained areas but was on the benches, desks, floors and in the air ducts. The workers were breathing in dust and carrying it around on their skin and clothing to other communal areas and their homes.

The first human experiments with plutonium were authorised in April 1945. Three studies were scheduled: one at the Chicago Met Lab, another by Joseph Hamilton's Group in Berkeley and a third by Stafford Warren at Oak Ridge. The first injection took place on 10 April 1945.

Ebb Cabe was a 55-year-old African American construction worker at the Oak Ridge Gaseous Diffusion Plant, who was travelling to work in the back of a car with his two brothers. Shortly after passing through the security gate they collided head-on with a heavy lorry. Everyone in the vehicle was taken to the Oak Ridge Army Hospital. Ebb's right forearm, his left femur and his right kneecap were fractured and he had cuts to his nose and lip. His injuries were not critical, but the driver Jesse Smith ended up in hospital for 9 months with a fractured hip. Ebb was able to give the doctors his family history and apart from a chronic urethral discharge he told them he was in good health. The medical examination also showed that he had a cataract in his left eye, marked tooth decay, gum disease, an arthritic left knee and reduced kidney function. Overall the doctors considered him a well-nourished and well-developed coloured male. A couple of days after the accident Wright Langham, the Los Alamos chemist, sent the medical team at Oak Ridge a sample of plutonium with instructions describing how to draw up and administer the injection without leakage and how to collect urine and faecal samples. As part of his medical care Ebb Cade was given a 0.24 milligram injection containing 4.7 milligrams of plutonium, an activity of 185,000 Bq. This was given without any prior explanation and without consent. Five days after the injection and nearly 3 weeks after the accident Ebb was taken to surgery for his bone fractures to be set. This timing allowed samples of bone to be taken to confirm the skeletal uptake of plutonium. When Ebb awoke 15 of his teeth and some bone from his jaw had been removed. The doctors had previously noted his gum disease and decayed teeth and probably took these samples on the basis of previous knowledge of radium uptake in the jaw. Although Ebb was the first subject to be injected, he was grouped with the Rochester subjects and subsequently given the study code HP-12. No one has since been able to confirm who actually gave the injection.

Arthur Hubbard was a white businessman from Austin, Texas, who at the age of 68 had developed an egg-sized lump on his chin in June 1944. This had been removed and the surrounding area treated with radiotherapy, but the tumour, a squamous cell carcinoma, was back within weeks and required further surgery and radiation therapy. A friend of his, who was a surgeon, recommended that he go to Billings Hospital in Chicago, which was only a short walk away from the Chicago Met Lab. Arthur had been in Billings Hospital for a few weeks when at 9:17 on the morning of 26 April 1945 he had no idea that he had just become the second person to be injected with plutonium. Following his injection of 6.5 milligrams of the radioactive metal, Arthur's urine and stool samples were collected for the analysis of bodily excretion. His code name was CHI-1, signifying that he was the first patient to be studied in Chicago.

The third patient to be injected was Albert Stevens who was a 58-year-old house painter who was admitted to Ward B of University of California Hospital in San Francisco with suspected stomach cancer. After a series of tests, the radiologist and surgeon concluded that he probably had gastric cancer but suggested a gastroscopy (placing a scope in the stomach) to confirm the diagnosis. The procedure was never carried out. Joseph Hamilton and his colleagues in Berkeley who had been undertaking animal studies were keen to find suitable patients and Albert seemed to be an ideal candidate. He was given the study code CAL-1 and became the first patient in California to be given plutonium. The injection was given on 14 May 1944 and he was taken to surgery 4 days later. During the operation the surgeons removed a mass but left his stomach virtually intact. They also removed his spleen, part of his liver and pancreas, some lymph nodes, abdominal fat and most of his ninth rib. The operation went well, and Albert returned to the ward in relatively good condition still unaware that he was the subject of 'top secret' medical research. Albert was unemployed and the cost of his care was rising. The hospital insisted that his family replace the blood transfusions he had needed, so his son and daughter flew in from

Michigan to give back the equivalent amount of blood that was used. After the operation each of the excised surgical samples were divided, with half going to a 'special study' and the remainder going to the path lab. Examination of the specimens was carried out by the hospital pathologist James Rinehart, who concluded that Albert had a benign gastric ulcer with chronic inflammation. This diagnosis was met with disbelief by the medical team, especially since Albert had undergone such extensive surgery with removal of various tissues and organs. Despite this he lived for a further 20 years. His medical record showed that he had signed consent for the anaesthetic and surgery but there was no record of any injection of radioactivity. Similar cases followed, as more patients were selected, all of whom were unaware that they had become guinea pigs in secret medical experiments. Between 1945 and 1947 a total of 18 patients were injected with plutonium by the Manhattan Project medical team. This included 11 patients at Rochester, New York, 2 patients at Billings Hospital, Chicago, and 3 patients at San Francisco. A summary of the medical details is shown in Table 7.1. Looking at the study details that are available it can be seen that the information was not always complete. It would appear that the patients received different amounts of activity and from the figures available it is not easy to correlate the activity injected with the radiation doses that each patient received nor is it obvious that the higher radiation doses resulted in a shorter survival time.

At the time, these experiments were seen as vital top-secret national research to gain knowledge of the potential effects of handling radioactive materials and the deployment of nuclear weapons. In total, researchers performed thousands of human experiments that were largely financed or supervised by the US military, the Atomic Energy Commission and the Federal Government. Many experiments were undertaken to determine the effects of radiation intake and radioactive contamination on the human body and were carried out unknowingly on poor, sick and vulnerable people. This included six patients at Rochester who were injected with varying amounts of uranium-234 and uranium-235 to study how much uranium their kidneys could tolerate before becoming damaged. Researchers at Vanderbilt University in Tennessee gave 829 pregnant mothers what they were told were vitamin drinks that would improve the health of their babies. The solution contained radioactive iron to determine how fast the radioactivity crossed the placenta from the maternal bloodstream to the foetus. It is thought that at least three children subsequently developed cancer such as leukaemia. In 1949, the Quaker Oats Company working with the National Institute of Health and the Atomic Energy Commission carried out an experiment by feeding minute doses of radioactive materials to boys at the Fernald School for the mentally retarded in Waltham, Massachusetts. This was to determine if the contents of the breakfast cereal prevented the body from absorbing iron and calcium. The boys were told that they were joining a science club, and the consent form that was sent to their parents gave no mention of the radiation used in the work.

Many other experiments were performed, including whole body irradiation on poor and black patients with cancer in Cincinnati, where doses of up to 1 Sv were given and irradiation of the testicles of prisoners in Washington to assess the effects of radiation on testicular function. The details of these and many other experiments remained classified for 50 years. It was not until the early 1990s that the *Albuquerque Tribune* uncovered the nature of the human experiments and also the

TABLE 7.1

Details of the 18 patients who were recruited to the first Manhattan Project plutonium study. Adapted from The Human Plutonium Injection Experiments by William Moss and Roger Eckhardt in *Radiation Protection and the Human Radiation Experiments*. Los Alamos Science Number 23, 1995. Editor N G Cooper

Case number and description	Clinical notes	Date injected	Survival time	Age at death (years)	Cause of death	Pu-239 activity injected (kBq)	Effective dose (Sv)
HP-12 55-year-old male	Motor accident victim at Oak Ridge Hospital; bone sample and teeth removed	10 April 1945	8 years	63	Heart failure	10.7	2.3
CHI-1 68-year-old male	Cancer of chin, metastasis to lungs; near death when injected; post-mortem samples taken	26 April 1945	5 months	68	Cancer of chin with lung metastases	14.8	0.2
CAL-1 58-year-old man	Gastric mass, misdiagnosed as stomach cancer. Tumour and other tissue taken at surgery	14 May 1945	20.7 years	79	Heart disease	1.7 (Pu-239) 130 (Pu-238)	64
HP-1 67-year-old man	Duodenal ulcer, with severe gastrointestinal bleeding	16 Oct 1945	14.2 years	81	Bronchopneumonia	10.4	3.8
HP-2 48-year-old man	Haemophilia and heart disease	23 Oct. 1945	2.4 years	50	Brain disease	11.5	0.8
HP-3 48-year-woman	Rash, hepatitis and hypoproteinaemia	37 Nov 1945	37.2 years	85	Acute cardiac arrest	11.1	8.8
HP-4 18-year-old woman	Metabolic disorder	27 Nov 1945	1.4 years	20	Cushing's syndrome	11.1	0.46
HP-5 56-year-old man	Lou Gehrig's disease. Sclerosis affecting nerves, brain and spinal cord	30 Nov 1945	4.9 months	57	Bronchopneumonia	11.5	0.14

(Continued)

TABLE 7.1
(Continued)

Case number and description	Clinical notes	Date injected	Survival time	Age at death (years)	Cause of death	Pu-239 activity injected (kBq)	Effective dose (Sv)
CHI-2 56-year-old woman	Breast cancer	27 Dec 1945	17 days	56	Metastatic breast cancer	218.3	0.3
CH-3 Young adult male	Hodgkin's disease	27 Dec 1945	5.6 months	Not known	Probably Hodgkin's disease	218.3	3.0
HP-6 44-year-old male	Addison's disease. Hormone deficiency	1 Feb 1946	38 years	82	Natural death	12.2	9.9
HP-7 59-year-old woman	Rheumatic heart disease	8 Feb 1946	8.5 months	60	Lung failure	14.4	0.3
HP-11 69-year-old man	Chronic malnutrition, alcoholism and liver cirrhosis	20 Feb 1946	6 days	69	Bronchopneumonia	14.8	0.01
HP-8 41-year-old woman	Scleroderma and duodenal ulcer	9 March 1946	29.7 years	71	Unknown	14.8	10
HP-9 64-year-old man	Dermatitis and general weakness	3 April 1946	1.2 years	65	Bronchopneumonia	14.4	0.5
CAL-2 4 years and 10 months boy	Bone cancer	26 April 1946	8.4 months	5	Osteosarcoma of bone	6.3	0.13
HP-10 52-year-old man	Acute heart failure	16 July 1946	10.9 years	63	Heart disease	14.1	4.1
CAL-3 36-year-old man	Suspected bone cancer of the left knee. Leg amputation and removal of bone tissue containing 50% of plutonium injection	18 July 1946	44 years	80	Lung failure	3.5	1.5

identity of the patients who were unknowingly enrolled into these studies. One of the journalists from the *Tribune* published a book in 1999 which was entitled *The Plutonium Files: America's secret medical experiments in the cold war.* In response to these revelations President Bill Clinton set up an Advisory Committee on Human Radiation Experiments in 1994. Most of the American population were unaware of these experiments until this investigation. According to the 1995 report the Advisory Committee concluded that there was no single case where there was any expectation that these patients would benefit medically from the injection of radioactivity. However, the committee found that human radiation experimentation during the period between 1944 and 1974 contributed significantly to advances in medicine and thus to the overall health of the public. With some exceptions, most of the human studies involved adult subjects and were 'tracer' studies involving small amounts of radioactivity that were unlikely to have caused physical harm. These experiments were carried out at a time when patients generally had complete faith in the doctors that were treating them. The doctors could do what they thought was best and their actions were rarely questioned. There is no doubt that by today's standards the conduct of the work and clinical oversight was entirely unethical. This has been the same for many medical discoveries of the past. When the English physician Edward Jenner tested his immunity hypothesis in 1796 by inoculating the 8-year-old son of his gardener, James Philips, with cowpox pus scraped from the hands of a milkmaid, this could hardly be considered to be ethical practice.

7.6 UK HUMAN TISSUE STUDIES

The controlled chain reaction in a nuclear reactor generates intense heat that can be used to produce steam to drive a turbine for electricity production. At 16 minutes past 12 on 17 October 1956, Queen Elisabeth II pulled a lever to switch on the first commercial nuclear power station in the world. Within seconds the nearby town of Workington, which had been built years earlier for coal and iron workers, was using heat, light and electrical power from nuclear fission. At the time, this source of cheap and clean energy was described as being the start of the second industrial revolution. Despite the huge nuclear programmes of America and Russia, the UK had beaten them to build and commission a power plant at Calder Hall, 15 miles south of the nuclear plant at Sellafield on the remote coast of Cumbria. The Sellafield site also included other nuclear reactors, including the Windscale Pile 1 reactor which caught fire within a year of the Queen's visit, causing the UK's worst nuclear accident.

In the same way that the US scientists wished to learn more about the human effects of working with radioactivity, the UK nuclear industry recognised that the production, use and reprocessing of radioactive materials exposed workers to unknown radiobiological hazards. From the start of their weapons and energy programmes, the United Kingdom Atomic Energy Authority and later British Nuclear Fuels Limited undertook secret research to monitor radiation exposure and radioactive uptake in their staff. Between 1960 and 1992 these organisations undertook radiochemical analysis of organs removed at post-mortem examination from 76 bodies of former workers. This included men who had worked at Sellafield and other UK nuclear sites, including Springfields, Capenhurst, Winfrith, Dounreay and Aldermaston. The workers and their families were unaware that a number of different organs including lungs, liver, kidneys, spleen, ribs, vertebrae and thigh bones were being removed. In April 2007 Michael Redfern QC was asked by the UK government to investigate the nature and extent of these undertakings. The enquiry subsequently went on to find that the bones from more than 6,000 randomly chosen people, many of them babies, had been removed to measure the levels of strontium-90 in studies that had been approved by the Medical Research Council. Although much of the research was carried out with the best of intention, the Redfern Report which was published in 2010 concluded that the pathologists involved had not followed the law by obtaining the informed consent of the individuals concerned or their families. As with US human studies these experiments were conducted according to the poorly defined medical ethics of the time.

Following centuries of physician autonomy, the ancient Hippocratic Oath was changed by international agreement in 1964 when the World Medical Association published the 'Declaration of Helsinki'. This required doctors to have the utmost respect for human life, ensuring that individual human welfare must always take precedence over the interests of science and society. The original version has been amended seven times since 1964, most recently at the General Assembly of the World Medical Association which took place in October 2013. The Declaration of Helsinki remains the cornerstone for the governance of medical research. Randomised controlled clinical trials have now superseded the clinical judgement of the individual physician. Thankfully, medical research in most countries is now regulated and governed by these agreed international harmonised research procedures and all clinical trials are approved by an independent ethical committee.

8 ☢

From Tracers to Treatment

A metallic chemical element, Like the colour of silver-grey.
Every isotope radioactive, Used for imaging like an x-ray.
Two physicists made this discovery, With excitement and delirium.
Artificial production accomplished, Designating it, Technetium!
Anonymous poem marking the discovery of technetium, the lost element by Emilio Segré and Carlo Perrier in 1937.

8.1 THE MARKED CARD

When the stakes are high, a cheating gambler can increase his winnings by marking his cards. This has been done many times by using subtle markings, invisible ink and other ingenious hidden indicators. In 2016, a waste treatment plant in Rüdersdorf to the east of Berlin detected some radioactivity in a rubbish truck during a routine checking procedure. Radiation monitoring of waste delivered to recycling centres is common practice in many countries to safeguard health and the environment. This was not the first time radioactivity had been found at a domestic waste plant near Berlin. Radioactivity had previously been found in 2014 at an incinerator plant in Brandenburg to the west of the city. Determined to look further into the second incidence, staff from the German Federal Office for Radiation Protection searched the rubbish bags at the Rüdersdorf plant using radiation monitors and found radioactivity in small fragments of playing cards. After further examination radioactivity was also found on waste food and cigarette ends. The police and radiation protection authority set about to determine if they could figure out where radioactive playing cards could have come from. They retraced the route of the waste truck to check if the locations of the waste collection fitted with the items of rubbish. The trail led to a Berlin restaurant, and after searching the premises they went on to the 41-year-old restaurant owner's home in Marzahn-Hellersdorf. After carrying out surveys with radiation monitors at both properties they recovered 13 playing cards marked with radioactive substance. The police deduced that the playing cards had been used in illegal card games, where the cards marked with radiation would only be identified by a player using a concealed detector placed under a sleeve or under clothing on their person.

The radioactivity turned out to be iodine-125 which has a half-life of 60 days and is commonly used in medical biochemistry units and university research laboratories. A total of 46 contaminated items were identified with a total activity of 28 MBq. The labelled playing card chips contained a total activity of 24.4 MBq. The maximum ambient dose rate was 12.6 mSv per hour at a distance of 10 centimetres from the back of the card. Detailed analysis of the gamma spectrum obtained showed that the iodine-125 was extremely pure, suggesting that it was obtained from material originally produced for medical use. It was subsequently estimated that the concentration of the iodine-125 solution used for marking the playing cards was in the range of 7 GBq to 20 GBq per millilitre. The amount of radioactivity recovered from the premises was not considered to be dangerous to health, although it could have resulted in some radioactive uptake in the thyroid glands of those handling the material. The German authorities ordered the contaminated areas to be sealed and cleaned but considered that there was no wider risk of contamination. The restaurant owner was charged for the environmental release of ionising radiation.

It was not clear what type of game the cards were being used for, but this was certainly an illegal undertaking, as the venue was not authorised for gambling. Low-energy gamma spectroscopic analysis using high-purity germanium radiation detectors indicated the presence of a thin layer of lead within the structure of the tagged playing cards. The lead inside the card allowed the cheating player to identify the face-up and face-down orientation of the card. This allows manipulation of gambling games such as the Vietnamese betting game 'Xóc Dĩa'. Both European-style poker cards and Asian-style playing cards are known to have been manipulated in this way. On 15 August 2014, police in Quang Ninh province of Vietnam arrested a man carrying four decks of playing cards that had been marked with radioactivity. The suspect, Bui Dinh Chung, confessed to police that a sensor concealed on his person generated vibrations when the marked cards were dealt his way. Chung was charged with stockpiling and trading in radioactive materials. The cards were believed to have been made in Taiwan, Hong Kong and mainland China and were reportedly used in illegal gambling dens. More recently, Romanian authorities identified an organised crime group that was involved in two incidents involving playing cards contaminated with iodine-125 when they were detected at a Bucharest airport in 2018. Using nuclear forensics skills and equipment with the support of the IAEA, the Romanian authorities initiated a criminal investigation that determined that the cards were being used to cheat at the popular *Xoc Dia,* card game.

It would seem that the use of radioactivity to cheat at gambling is not restricted to card games. There are reports of a portal detector being triggered during a routine check at Chinggis Khan international airport in Mongolia, indicating the presence of a radioactive substance in the bag of an incoming passenger. The radioactivity was found on three gaming dice. On each one, the side with four dots was painted with paint containing iodine-125. The dose rate measure from the three dice was greater than 2 mSv/h near the surface. In April 2010 a number of passengers were ordered to leave the terminal at Xiamen Wutong Ferry Port in China when radioactivity was found in a passenger's luggage. According to the Xiamen Inspection and Quarantine Bureau the dice were coated with americium-241 which is commonly used in domestic smoke detectors. A further recorded incident occurred on 5

April 2014, when Chomrong Bun of Columbus, Ohio, arrived at Atlanta Hartsfield Jackson International Airport on a Korean Air flight from Incheon, Korea. During immigration, a radiation detector alerted the immigration officers of the presence of radioactivity in her luggage. Inspection revealed several items, including radioactive dice and a radiation detection device. She was charged in June 2015 at the District Court of Atlanta, Georgia, with recklessly transporting radioactive iodine-125 on a commercial passenger flight.

8.2 ATOMIC GOLF BALLS

It is not surprising that there have been some imaginative ways of using radioactivity to follow things around. Soon after the Second World War, an era when nuclear-powered trains and planes were seriously being considered, the Ford motor company was making designs for the 'Ford Nucleon', a car powered by a small nuclear reactor behind the back seat and would travel 5,000 miles between recharges. In 1951, Dr William Davidson of the BF Goodrich tyre company came up with the idea of the golf ball that couldn't get lost. By embedding small amounts of radioactivity under the surface of the golf ball, Davidson claimed that it could always be found even after a slice hit into the deepest of undergrowth by using a hand-portable Geiger counter and headphones. The atomic golf ball never made it to market, since the gamma rays from the small radioactive source were of low energy and so the golfer had to be within a few feet to locate the lost ball. The prospect of carrying a Geiger counter costing $25 was also a deterrent when you could buy over 25 new balls for the same amount of money. Years later the Canadian Atomic Energy Labs did make a claim that irradiating golf balls with high-energy electrons altered the ball's inner molecular structure which would increase the amount of energy absorbed per hit, adding 20 to 30 yards to a drive stroke. This claim was never proven.

8.3 TRACK AND TRACE

The concept of using small amounts of radioactivity as 'tracers' to mark and track the movement of materials from one place to another goes back to the early 1900s. In January 1911 George de Hevesy, a Hungarian chemist, left his position at Karlsruhe in Germany and crossed the English Channel to take up a position with Ernest Rutherford at Manchester University. Rutherford had been working on the decay products of radium which were called radium A, radium B, radium C, etc. He had obtained a large amount of radium D which was of considerable value, but the samples seemed to be contaminated with a large amount of lead. Rutherford thought that separating the lead from the radium was a problem that this bright young Hungarian chemist could solve. Hevesy eagerly took on the task, but he failed completely, since radium D finally turned out to be a form of lead, lead-210 in fact. Hevesy became frustrated and grew distinctly unhappy with the work. On top of this he also developed a dislike to the Manchester weather, the boarding house where he was staying and the food the landlady was serving him. It may have been bouts of indigestion that made him more disgruntled but he became convinced that his landlady had the unappetising habit of recycling his leftover food. When he asked her to serve freshly prepared meat more than once a week, she was indignant and insisted that her cooking used only the freshest of ingredients but Hevesy wasn't convinced. At the end of one of his Sunday roast dinners, Hevesy secretly spiked the leftover meat on his plate with some radioactivity taken from his laboratory. When sitting down for his meal a few days later, he placed an electroscope on the table which revealed the presence of the spiked radioactive leftovers on his plate of recycled hash. This first radiotracer investigation had followed the leftover meat from Hevesy's Sunday dinner plate to the kitchen meat grinder, into the stew pot and back onto the dining table days later. His landlady had to admit to the deed claiming that he was using some kind of magic. Hevesy's real scientific brilliance followed when he considered how he could use this technique in his work. He took a known amount of pure radium D that had decayed from radon to study the properties of lead. He left for Vienna where he knew he could obtain pure radium D and by 1913 published the first scientific account of the radiotracer technique. This was the beginning of what was to be an extremely important new branch of science. On a previous occasion, whilst working at the Manchester laboratory Hevesy was having a cup of tea with the English physicist Henry Moseley. As they drank together, they considered the process of tracking the water in the tea as it passed through the various organs and tissues of the body, before finally being excreted in urine. Henry Moseley is remembered for his work in X-ray spectroscopy, but he never took part in the tea experiment. During the First World War he volunteered as a telecommunications officer in the British army and was killed by a bullet in the Battle of Gallipoli in Turkey in 1915. Hevesy was later given the opportunity to follow water molecules through the body in 1933 after the American physical chemist, Harold Urey, had completed his work on identifying heavy water (deuterium oxide). Deuterium is not radioactive but can be identified by optical spectroscopy. Urey presented Hevesy with a few litres of deuterium, so he could test out his theory. Hevesy found a new drinking partner and after drinking measured amounts of heavy water they recovered over 55 samples of urine and other excreta, allowing them to estimate that the body contained 43 litres of water. This important new science led to the increased understanding of metabolic function and basic physiology and started a new era of biomedical investigation. Hevesy went on to use natural radioactive tracers such as lead-210 (radium D) and lead-212 (thorium B) to study the uptake of lead in the roots, stems and leaves of plants. He then went on to look at the metabolism of lead in animals, and his research really took off when the newly developed methods for the production of artificial radioactivity gave him new elements to investigate. He went on to use phosphorus-32, sodium-24 and potassium-42 for extensive studies of metabolic and physiological processes and he was awarded the Nobel Prize in Chemistry in 1943 in recognition for his work. George de Hevesy is widely acknowledged to be 'the father of nuclear medicine'. A postage stamp shown in Figure 8.1 was printed in his home country in 1988 to commemorate his work.

One other story has to be retold of George de Hevesy's ingenuity. After the peace activist Carl von Ossietzky had received the Nobel Peace Prize in 1935, the German government prohibited citizens from accepting or keeping any Nobel Prize awards. When Germany invaded Denmark, the Nazis intended to confiscate the Nobel gold medals of the German physicists Max von Laue (1914 recipient) and James Franck (1925 recipient), which had been sent to Niels Bohr's Institute of Theoretical Physics

FIGURE 8.1   Hungarian stamp printed in 1988 commemorating the work of George de Hevesy.

Source: Hungarian stamp printed in1988 commemorating the work of George de Hevesy. www.shutterstock.com/image-photo/hungary-circa-1988-stamp-printed-by-90600943

in Copenhagen for protection. Hevesy was working with Bohr at the time and they discussed how to protect the medals from being stolen. One possibility was to bury them, but Hevesy decided to dissolve the gold in nitro-hydrochloric acid and then placed the resulting orange solution in a flask on a shelf in the laboratory. After the war, he returned to find the solution undisturbed and precipitated the gold out of the acid. The gold was then sent back to the Nobel Foundation in Stockholm, where the medals were recast. They were presented back to their owners in 1952.

8.4 THE NUCLEAR CLOUD'S SILVER LINING

In 1935 Robert Oppenheimer attended a meeting at Berkeley in California with Ernest Lawrence, one of his colleagues who had invented the cyclotron in 1929 (Figure 8.2). The cyclotron was a high-energy particle accelerator that at the time was making a string of new discoveries with the production of artificial radioisotopes. In front of the assembled group Lawrence placed Oppenheimer's hand around a Geiger counter which was producing the occasional random click in response to the normal level of surrounding background radiation. He then gave him a glass of water to drink. After the glass was empty everything was quiet for around half a minute, but after about 50 seconds the Geiger counter kicked into life with rapid and sustained clicking sounds. The water had contained a small amount of radioactive sodium-24, and Lawrence's simple demonstration had shown the complex physiochemical process of a salt (sodium chloride) taken up from Oppenheimer's gastrointestinal tract into his bloodstream and circulating through his veins and arteries to reach the tip of his fingers within a minute. This simple, impromptu demonstration in front of an amused audience showed the remarkable ability of a small amount of radioactivity to reveal the physiological function of the human body. Whilst this showed the potential of an important new branch of medical science, from today's perspective the fact that Lawrence performed this experiment without asking Oppenheimer's permission or even mentioning that the water contained radioactive sodium is perhaps questionable. Lawrence knew from prior research that the experiment was not dangerous, and his colleague was equally happy that this small amount of radioactivity would not cause him any lasting harm (Figure 8.2).

FIGURE 8.2   Robert Oppenheimer and Ernest Lawrence climbing down from the 184-inch cyclotron at the Radiation Laboratory, University of California, Berkeley in 1946.

Source: Robert Oppenheimer and Ernest Lawrence climbing down from the 184 inch cyclotron. www. shutterstock.com/editorial/image-editorial/ historical-collection-10293008a

There is no question that the science of the Manhattan project was brilliant. It had built on earlier knowledge in physics and biology and established methods for the medical use of radioactive materials in the human body. The radiotracer studies performed at Los Alamos, although initially motivated by radiation safety concerns, made a huge contribution to the understanding of human biology and medicine. The safety limits established at Los Alamos for the amounts of radionuclides within the human body enabled doctors to safely administer radioisotopes to humans for research, diagnosis and therapy. Although there was no excuse for the way the human radiation experiments of the time were undertaken, it is notable that the scientists and medics working with these new radioactive materials undertook many of the first experiments on themselves. One notable individual was Wright Langham, an analytical chemist who worked with the medical team at Los Alamos. His contribution to the science of radiobiology is often overlooked because he was so closely involved with the human plutonium studies. He pioneered the Los Alamos work for the assay of plutonium in urine samples and used whole body radiation counting equipment to determine retained activity in the body. He also developed methods for estimating human radiation doses using test objects filled with radioactivity to mimic the body and its internal organs. He was often seen with the plastic men he used in his experiments (Figures 8.3 and 8.4). These test objects are now known as 'phantoms' and are widely used by medical physicists and medical equipment engineers for the calibration of X-ray machines, gamma cameras, PET scanners and radiotherapy units. Tragically Wright Langham died along with eight other passengers in a plane crash in 1972 when an aircraft shuttle to Los Alamos, chartered by the Los Alamos Scientific Laboratories, lost power and nose-dived into a field shortly after taking off from Albuquerque International Airport.

8.5 NUCLEAR MEDICINE

During the 1930s, the cyclotron at Berkeley started to produce new artificial radionuclides such as sodium-24, potassium-42, iodine-128, iron-59, chlorine-34, phosphorus-32 and bromine-82 that proved to be of vital importance in biomedical science. Sodium-24 was used to measure the circulatory time of the blood and iron-59 for the first time established the kinetics of iron uptake and storage in bone marrow and its rate of incorporation in red blood cells (human ferrokinetics). Phosporus-32 was used as a successful treatment for polycythaemia vera, a disease where the blood marrow produces too many red blood cells, causing high blood pressure, severe headaches and leading to life threatening blood clotting.

The development of nuclear medicine as a clinical speciality was largely based on the use of radioiodine. Iodine with an atomic number of 53 is the heaviest element required for human metabolism. All the iodine taken into the body is taken up by the thyroid gland which is a butterfly-shaped organ in the neck. Iodine is essential for production of the thyroid hormones, chiefly triiodothyronine and thyroxine which help regulate cellular metabolism, the speed at which chemical reactions take place in the body. To maintain health, adults need to consume around 140 micrograms of iodine a day. The main dietary sources of iodine are sea fish and shellfish, but it can also be found in cereals and grains depending on the amount of

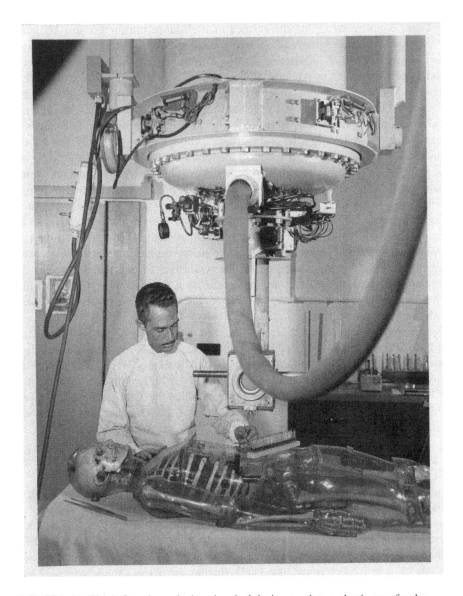

FIGURE 8.3 Wright Langham placing chemical dosimeters into a plastic man for the study of biological radiation effect at Los Alamos. The plastic men could be filled with radioactive materials simulating the uptake in human organs.

Source: Wright H Langham, Assistant Health Division Leader for Bio-Medical Research. Photograph reproduced courtesy of the University Archives, Library Special Collections, Charles E Young Research Library, UCLA

iodine in the soil where the crops are grown. In inland populations, living away from the sea iodine deficiency can lead to thyroid diseases. If the thyroid is underactive, the condition is known as hypothyroidism. This is mostly caused by an autoimmune reaction when the immune system which usually fights infection attacks the

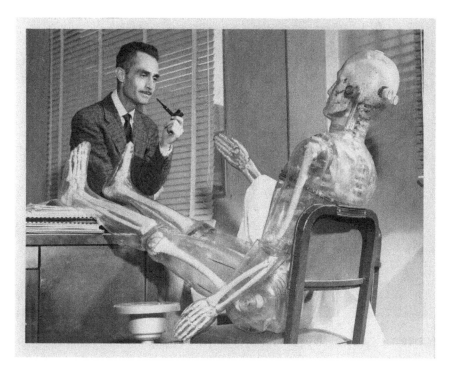

FIGURE 8.4   Wright H Langham, Assistant Health Division Leader for Bio-Medical Research, and one of the two plastic men used in the studies of radiation effects on humans.

Source: Wright Langham placing chemical dosimeters into a plastic man. Photograph reproduced courtesy University Archives, Library Special Collections, Charles E Young Research Library, UCLA

thyroid. In hypothyroidism the gland does not produce enough thyroid hormones. The symptoms include tiredness, depression, dry skin and hair and slow movement and thoughts. If the thyroid is overactive this is known as hyperthyroidism. One particular characteristic of hypothyroidism is an enlargement of the thyroid gland causing swelling in the neck (Figure 8.5). This condition was recognised by physicians in the 18th and 19th centuries, in people living in the central English county of Derbyshire (giving rise to the condition of Derbyshire Neck or Derbyshire Goitre) and also in the central European area of the Swiss Alps and the Great Lakes and Midwestern regions of the US.

Despite the introduction of iodine-enriched salt, thyroid disease remains a major health problem in some populations, particularly in the central regions of the Indian continent. Patients with an overactive thyroid have a number of miserable symptoms in addition to neck swelling. The symptoms include nervousness, anxiety, irritability twitching and trembling. Patients often have an irregular or unusually fast heart rate (palpitations), have difficulty sleeping, feel persistently weak and tired and have mood swings.

The body is unable to differentiate between radioactive and non-radioactive iodine and so they are both metabolised in an identical manner. Substitution of radioactive atoms of iodine for naturally occurring atoms provides a means of monitoring

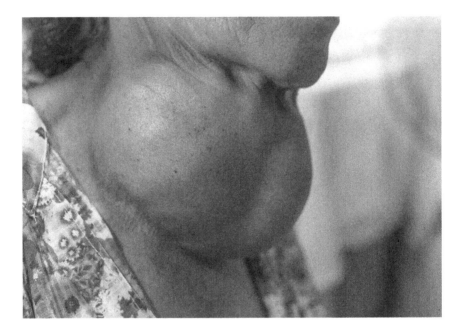

FIGURE 8.5   A neck goitre in an elderly Asian female patient.

Source: Neck goitre in an elderly Asian female patient. Shutterstock, www.shutterstock.com/image-photo/old-asian-female-very-large-thyroid-655176013

the iodine distribution in the body and can be used to measure iodine metabolism in the thyroid gland. At Berkeley artificially produced iodine-131 was used by Dr John Lawrence (the brother of Ernest Lawrence) and Dr Joseph Hamilton to examine human thyroid function. A diagnostic test was developed which simply required the patient to drink a glass of water containing iodine-131 which emits both gamma rays and beta particles. The iodine is absorbed from the gut into the bloodstream, where it is taken out of the circulation almost exclusively by the thyroid gland. By holding a Geiger counter over the thyroid, he could measure the amount of iodine-131 taken up by the gland. If the thyroid took up too much, the diagnosis was hyperthyroidism, whereas too little uptake could mean hypothyroidism. With the subsequent development of scanners and gamma cameras a thyroid scan could be undertaken to identify regional uptake which can be shown as hot and cold areas of iodine uptake in the thyroid. This can help with planning appropriate treatment, such as surgery. Whilst hyperthyroidism can be treated by surgery or with medicines, the simplest way of treating the patient is to give patients a small drink of iodine-131. The beta particles that are emitted from the iodine deliver a sterilising amount of radiation targeted directly to the thyroid tissue, killing the overactive cells. This type of internal radiotherapy treatment is convenient and highly effective and has been used to improve the lives of many millions of patients throughout the world.

Scientific and technological advances in nuclear methods over the past 70 years have resulted in an increased range of radionuclides suitable for medical use. There are now three main production routes for the manufacture of medical radioactive products. In addition to cyclotron bombardment with charged particles (usually protons), a nuclear reactor can be used to irradiate materials with neutrons (neutron activation) or to irradiate uranium targets causing them to undergo fission. The final method requires chemical reactions to separate out the required fission products, a rather messy process to undertake with highly radioactive materials.

8.6 TECHNETIUM, THE MISSING ELEMENT

There was one particular element which had remained elusive since Dmitri Mendeleev first produced the periodic table. This was element number 43, which should have occupied the space between element 42, molybdenum, and element 44, ruthenium. Mendeleev had given it the name 'ekamanganese' since 'eka' meant one in Sanskrit, signifying that it was one place down from manganese in column 7 of the table. In 1936 Emilio Segrè, an Italian American physicist, visited the cyclotron facility at Berkeley and persuaded Ernest Lawrence to let him have some discarded parts from the cyclotron runs that had become radioactive. Lawrence sent him a piece of molybdenum metal foil. Working with his colleague Carlo Perrier at the University of Palermo in southern Italy, Segrè succeeded in isolating the elusive element when he identified two radioactive products: technetium-95m and technetium-97. The 'm' in '95m' was designated a short-lived energetic level known as a 'metastable state' that quickly decayed giving off gamma rays. The university authorities wanted to name the new element 'panormium' after 'Panormus', the Latin name for Palermo (the name of the university); however, it was finally given the name 'technetium' after the Greek word 'technos' meaning artificial. Segrè returned to Berkeley and met Glenn Seaborg and together they isolated the metastable radionuclide technetium-99m. This radioactive metal had a physical half-life of 6 hours and was a pure emitter of gamma rays having an energy of 140 keV, this being near the upper energy level of medical X-rays and ideal for gamma imaging. The other convenient aspect of technetium-99m is that it is the decay product or 'child' of molybdenum-99, a fission product from uranium with a half-life of 66 hours. In 1947 the Brookhaven National Laboratory was established on the former site of an army camp in Upton New Jersey at the eastern end of Long Island to explore the peaceful use of radioactivity. By 1958 researchers Walt Tucker, Powell Richards and Margaret Greene had developed a way to bind the molybdenum-99 parent to aluminium oxide and separate the technetium-99m child decay product by the simple process of washing with salt solution. The system was originally called an 'isotope cow' as the process was originally likened to milking a cow.

Technetium plays no natural biological role and is not normally found in the human body, but it can be chemically bound to a variety of compounds which are of value in medicine. The first known medical imaging study with technetium-99m was carried out by Leif Sorensen and Maureek Archambault, who in 1963 reported the nuclear imaging work they had carried out at the Departments of Medicine and Radiology at the University of Chicago. In an authorised clinical trial they injected cancer patients with 1.5 MBq of molybdenum-99 and waited for 24 hours, so that sufficient decay had occurred to allow imaging with the gamma rays from the internally generated technetium-99m. Following injection into the bloodstream this tracer accumulated in normal liver tissue, demonstrating abnormal defects of uptake due to tumours and abscesses in the liver. The main problem with this approach was that the patient was exposed to the undesirable radiation from both the molybdenum and technetium decay. A number of clinical reports on medical imaging following the injection of technetium-99m followed, including one published in 1965 by a clinical group at the Liverpool Radium Institute in England using a technetium generator supplied by the Radiochemical Centre at Amersham, UK.[1] This work showed that technetium could be used for imaging the stomach and the thyroid gland and measuring blood clearance and radiation dosimetry. Other studies followed, showing that technetium-99m could be used for a wide range of diagnostic imaging investigations including brain, bones, heart, kidneys and lungs. Furthermore it was shown that the injection of technetium-99m for medical imaging did not expose the patient to any greater radiation exposure than was received from X-ray examinations.

After further development in the early 1960s, the technetium generator became more widely available and the clinical applications increased. In 1966 US production was transferred from Brookhaven to a number of commercial manufacturers. The current generator can be likened to a filter coffee machine, since it is of similar size and works in a similar way. In a coffee maker, when water is passed through the coffee grounds, the coffee is collected at the bottom and can be drunk free from the bitter grounds, which remain in the filter. In the technetium generator, passing sterile saline solution over molybdenum-99, a process known as elution, washes the technetium off in the form of technetium-99m pertechnetate whilst molybdenum remains bound to the column. Today most clinical nuclear medicine departments receive delivery of a technetium generator to the hospital on a weekly basis. The molybdenum decays with a half-life of 66 hours, allowing the generator to provide technetium-99m for daily clinical use over a period of 1 or 2 weeks. The technetium is then used to label a range of different tracer molecules to carry out various medical imaging examinations. The final sterile product given to the patient is known as a radiopharmaceutical and can demonstrate physiological processes in the body. A radiopharmaceutical contains two principal components: the radionuclide and the pharmaceutical compound. The radionuclide provides the means of detecting and imaging via the gamma rays (or of delivering a dose of radiation for therapy using beta or alpha particles) whilst the pharmaceutical molecule directs the tracer to the site of uptake in the required organ in the body. Since these are given as tracers, the amount of radioactivity is kept low and the amount of the drug or targeting compound is miniscule. Adverse reactions from receiving these pharmaceuticals are extremely rare. In addition to

---

[1] The Radiochemical Centre was established in 1940 when a company called Thorium Ltd was set up at Chilcote House in Amersham, Buckinghamshire, UK, for the extraction of radium to make luminous dials and instruments. Years later the Radiochemical Centre became Amersham International and was one of the largest producers of medical radiopharmaceuticals in the world. It was the first national organisation in the UK to be privatised by Margaret Thatcher's government in 1982 and later became part of GE Healthcare.

technetium-99m imaging is commonly undertaken with a range of radiopharmaceuticals labelled with iodine-123, indium-111 and thallium-201.

Early nuclear imaging was carried out by moving a radiation detector forward and back over the body giving rise to an image of dots, which became known as a 'gamma scan'. In 1957 Hal Anger, an electrical engineer working at the Donner Laboratory at Berkeley, invented a mapping device using a scintillation crystal which produced a light flash or scintillation for each individual gamma ray that was detected. The light flashes were detected by photocells and the position of each light pulse was recorded giving rise to an image. The device which was originally called the Anger camera later became known as a gamma camera. An example clinical image recorded by a gamma camera is shown in Figure 8.6.

Anger filed for a patent which was granted in 1961. He received considerable royalties from his invention which was used throughout the world for medical imaging. The basic principle of operation remains integral to the operation of modern medical gamma cameras today; however, most modern instruments have dual heads that can scan over and around the patient, producing whole body images and three-dimensional reconstructed slices, a process known as SPECT (single photon emission computed tomography). The combination of SPECT and X-ray-computed tomography (SPECT–CT) produces nuclear functional images overlayed onto the X-ray anatomical image. A common nuclear medicine procedure is a technetium-99m bone scan (Figure 8.7).

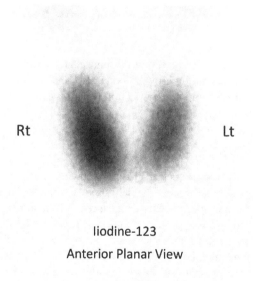

Rt   Lt

Iiodine-123
Anterior Planar View

FIGURE 8.6 Thyroid scan: a gamma camera image of the neck region of a 44-year-old male patient following injection of 13 MBq of iodine-123-sodium iodide. The image is recorded from the front of the patient, so the right lobe of the thyroid which shows higher uptake is on the left.

Source: Thyroid scan. Author's original material

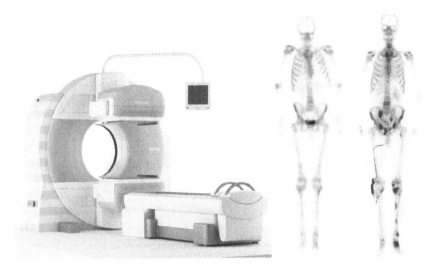

FIGURE 8.7   Dual head whole body gamma camera and whole body bone scans recorded 2.5 hours after the patient received an injection of a 400 MBq technctium-99m-labelled bisphosphate radiopharmaceutical. The left scan is normal. The right scan shows multiple hot spots of increased uptake due to bone metastases from bladder cancer. As well as uptake in the skeleton, activity can be seen in the urinary bladder and in the kidneys. The patient on the right has a urinary catheter draining into a collection bag at the knee.

Source: Dual head whole body gamma camera and whole body bone scans. Gamma camera image provided by Siemens Medical UK. Bone scan provided by the author

A single injection with 400 MBq of Tc-99m-bisphosphate is used to image the entire skeleton. The tracer is taken up by growing bone cells and shows areas of increased bone turnover due to bone tumours or infection. This procedure would result in an effective dose of 3 mSv.

8.7 THE ANTIMATTER SCAN

In recent years there is a type of nuclear imaging known as positron emission tomography (PET scanning) that has become widely used in clinical diagnosis, especially in the diagnosis and follow up of patients with cancer. PET requires the injection of a short-lived positron-emitting tracer such as fluorine-18, carbon-11, nitrogen-13 or zirconium-89. The positron is a positively charged electron which is the antiparticle (or antimatter) opposite number of the electron. Although they are strongly attracted to each other, when the two particles get together they literally go out with a bang and annihilate each other, producing a pair of gamma rays each travelling in opposite directions with energy of 511 keV. By placing a ring of scintillation detectors around the patient the two gamma rays can be detected and the position registered. The final image, containing the sum of all the coincident pairs of detected gamma rays is reconstructed by computer. Scientists from Duke University in North Carolina and Massachusetts General Hospital in Boston independently published the first studies of coincidence detection in 1951, and the first imaging device was developed by Brownell and Arnow in 1953. Since those early investigations, advances in electrical engineering and the introduction of computer technology have made a vast improvement in the quality of the images that can be produced. This has greatly increased our understanding of human physiology, helped in the development of new drugs and led to improved diagnosis and treatment. In 1984 a team led by Prof. Nagai working at Gunma University in Japan set up a side-by-side PET scanner and X-ray CT scanner, so that combined PET and CT images could be produced with patient laying on a single scanning couch. This work was not published and so it was not until 7 years later when the British medical physicist David Townsend and an American electrical engineer Ron Nutt first formally proposed combining PET and CT imaging in a single machine. They began clinical evaluation of the first PET–CT system at the University of Pittsburgh Medical Centre in 1998. By fusing functional nuclear medicine images with anatomical CT images, they dramatically improved the diagnostic information that could be gained from the scans, making them much easier for surgeons and physicians to understand and so leading to improved patient diagnosis and treatment. Since 2001 all PET scanners used in hospitals have been combined with CT scanners for acquiring co-registered hybrid anatomical and functional images. PET–CT is now regarded as the most specific and sensitive means for imaging molecular interactions and pathways within the human body. Although PET–CT is expensive to install and operate it has been demonstrated to be a cost-effective way of helping doctors treat their patients, reducing the cost of unnecessary surgery and inappropriate drug treatments. The most commonly performed PET scanning procedure uses an injection of a fluorine-18 labelled sugar molecule: F-18-FDG (fluorodeoxyglucose). This tracer is taken up in cells with high metabolic glucose turnover such as brain, brown fat and malignant tumours which are actively growing at a faster rate than the surrounding normal tissue (Figure 8.8).

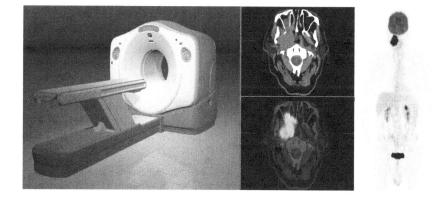

FIGURE 8.8   Left: a PET–CT scanner. Middle: combined PET–CT study; top: X-ray slice through the neck; bottom: fluorine-18-FDG PET slice at the same position, showing intense uptake in a tumour at the right side of the neck (images viewed from the feet). Right: fluorine-18-FDG PET image of the head and body showing intense uptake in a neck tumour. Normal physiological uptake of the tracer can be seen in the brain, liver, kidneys and urinary bladder.

Source: PET-CT scanner and Combined PET-CT study. PET scanner image provided by Siemens Medical UK. Bone scan provided by the author

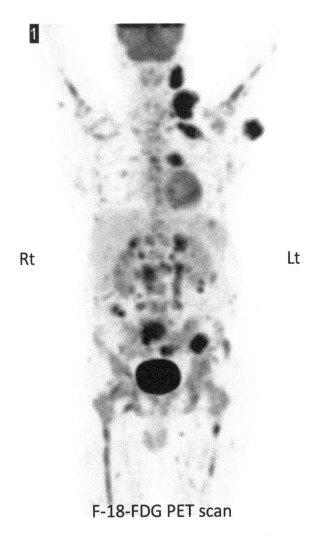

F-18-FDG PET scan

FIGURE 8.9 Fluorine-18-FDG PET scan of a patient with metastatic melanoma. Normal physiological uptake can be seen in the brain, heart and urinary bladder. All the other dark black 'hot spots' show widespread tumour uptake of the tracer in bone marrow, soft tissues and lymph nodes.

Source: Fluorine-18-FDG PET scan of a patient with metastatic melanoma. Image kindly provided by Dr Mboyo Di Tamba Willy Vangu, Head of Nuclear Medicine University of the Witwatersrand, Johannesburg Academic Hospital, South Africa

A further example of the value of PET imaging is shown in Figure 8.9. This was the scan of a patient who went to see his family doctor in Johannesburg, South Africa, after observing a small brown mole on the skin of his forehead. The doctor carried out a minor operation to remove the skin lesion and the patient went home thinking all was well. Unfortunately, the doctor forgot to send the excised tissue specimen for histological examination and the patient was never followed up. A year later he came back to his doctor feeling unwell. After some blood tests he was sent for a PET scan which showed the spread of the tumour throughout his body. By this time the disease was so extensive that treatment was ineffective and the patient subsequently died of metastatic melanoma.

Some patients may feel uncomfortable about the risks associated with receiving an injection of a radiopharmaceutical when undergoing a medical procedure, as they think this may be more harmful than having an X-ray. However, in general the radiation doses to patients undergoing nuclear medicine investigations are no greater than the doses from X-ray investigations. The doses received from X-ray CT examinations generally range between 2 and 18 mSv. Nuclear medicine imaging studies range from about 0.2 mSv for a lung ventilation investigation with a technetium-99m inhaled particulate to around 20 mSv for examination of the blood supply to the muscle of the heart using thallium-201-thallous chloride. An injection of 400 MBq of F-18-FDG for a PET scan results in an effective dose of around 7.6 mSv, but this will be higher if X-ray CT is also undertaken at the same time. An overriding consideration when assessing the potential harm from all medical radiation exposures is that the benefit gained from the diagnostic investigation is always greater than the radiation risk of carrying out the procedure and that the risk to the patient would be even higher if the investigation was not undertaken.

9

A Series of Unfortunate Events

The accident at the Chernobyl power plant in 1986 affected hundreds of thou-sands of people and was the most serious in the history of the nuclear industry. This showed that radiation release was not restricted to borders and led to improved cooperation between countries.

9.1 ACCIDENTS HAPPEN

In July 1957 the International Atomic Energy Agency (IAEA) was set up as an indepen-dent body to pursue the safe, secure and peaceful uses of nuclear sciences and technol-ogy. Based in Vienna, the IAEA reports to the United Nations General Assembly and the United Nations Security Council. In 1990 the IAEA introduced the International Nuclear and Radiological Event Scale to communicate the severity of incidents involv-ing radioactivity and radiation. This was intentionally designed to be similar to the Richter Magnitude Scale used to describe earthquakes. The nuclear accident event scale is logarithmic over 7 levels, with each increasing level representing an event approximately 10 times more severe than the previous level. Levels 1–3 are called 'incidents' and Levels 4–7 'accidents'. To date there have only been two Level 7 acci-dents: the Chernobyl and Fukushima disasters. Events without safety significance are called 'deviations' and are classified at Level 0. Table 9.1 shows the scale of nuclear and radiological events with examples of some of the major incidents that have occurred.

There have been many documented nuclear and radiological events, some of which have had both national and international consequences. The following events have been given greater detail because of their significance and historic interest.

9.2 THE WINDSCALE FIRE, 1957

The Windscale Piles were two air-cooled graphite-moderated reactors built on the site now known as Sellafield, on the coast of Cumberland in northwest England. These were constructed to produce weapons-grade plutonium for the British atomic bomb project. The accident occurred on 8 October 1957 when routine heating of the No 1 reactor went out of control leading to a fire that ruptured uranium cartridges in the reactor core. The deputy general manager at the Windscale and Calder works was a 39-year-old chemist Thomas Tuohy. When the fire started, he was at home about a mile

TABLE 9.1

IAEA classification of some major international nuclear and radiological events

Level	Classification	Summary description	Examples
7	Major accident	A major release of radioactive material with widespread health and environmental effects requiring planned extended countermeasures	**Chernobyl disaster, USSR, 26 April 1986**: a delayed test of the cooling system on the RBMK-type No. 4 reactor at the Chernobyl power plant was undertaken with lack of communication between staff and inadequate safety precautions, resulting in a criticality event that was the worst nuclear disaster in world history **Fukushima Daiichi Nuclear Disaster, Japan, 11 March 2011**: a series of events caused by the 2011 Tōhoku earthquake and tsunami caused damage to the backup power and containment systems resulting in overheating and radioactive leaks from the Fukushima I nuclear plant reactors. A temporary exclusion zone of 20 kilometres was established around the plant
6	Serious accident	A significant release of radioactive material with widespread health and environmental effects requiring planned counter measures	**Kyshtym Soviet Union Disaster, 29 September 1957**: a failed cooling system at a military nuclear waste reprocessing facility caused an explosion with a force equivalent to 70–100 tons of TNT
5	Accident with wider consequences	Limited release of radioactivity, likely to require some planned countermeasures Several deaths from radiation exposure	**Windscale fire, Sellafield, UK 10 October 1957**: heat damage to the moderator of a military air-cooled reactor caused a fire in the graphite and uranium fuel, releasing radioactive material as dust into the environment **Three Mile Island accident, Harrisburg, Pennsylvania US, 28 March 1979**: a combination of design and operator errors resulted in the gradual loss of coolant, leading to a partial meltdown and release of radioactive gas into the atmosphere **Goiânia incident, Brazil, 13 September 1987**: a forgotten caesium-137 radiotherapy source left in an abandoned hospital was removed by thieves who were unaware of its radioactive nature. Four people died and 249 people were contaminated

Level	Classification	Summary description	Examples
4	Accident with local consequences	Minor release of radioactivity unlikely to require planned countermeasures other than local food controls Damage or melting of fuel resulting in more than 0.1% release of the core radioactivity At least one death from radiation exposure	**Sellafield, UK**: five incidents involving release of radioactivity from 1955 to 1979 with no fatalities **SL-1 Experimental Power Station, US, 1961**: prompt criticality accident, killing three operators **Saint-Laurent Nuclear Power Plant, France: 1969**, partial core meltdown; **1980**, graphite overheating **RA-2 Research Reactor, Buenos Aires, Argentina, 1983**: a criticality accident during fuel rod rearrangement killed one operator and injured two others **Jaslovské Bohunice Reactor, Czechoslovakia, 1977**: partial core meltdown resulted in minor release of radiation to reactor building **Tokaimura nuclear accident, Japan, 1999**: three inexperienced operators at a uranium processing facility caused a criticality accident. Two of them died
3	Serious accident	Near-accident at a nuclear power plant with no remaining safety provisions Lost or stolen highly radioactive sealed source Exposure rates of more than 1 Sv/h in an operating area	**HORP Plant, Sellafield UK, 2005**: very large leak of a highly radioactive solution held within containment **Paks Nuclear Power Plant, Hungary, 2003**: damage to fuel rods whilst in cleaning tank **Vandellòs I Nuclear Incident, Vandellòs, Spain, 1989**: reactor shut down due to fire destroying control systems **Davis-Besse Nuclear Power Station, US, 2002**: inadequate inspections missing corrosion through the carbon steel reactor head leaving only 9.5 millimetre thickness of stainless steel cladding holding high-pressure reactor coolant
2	Incident	Significant failures in safety provisions but with no actual consequences. Significant contamination within the facility into an area not expected by design	**Sellafield 1, UK, 2018**: due to cold weather, a pipe failed causing water from the contaminated basement to flow into a concrete compound which subsequently discharged radioactivity into the Irish Sea **Hunterston B nuclear power station, Ayrshire, UK, 2018**: inspection revealed cracks in the graphite bricks of the advanced gas-cooled reactor 3

(Continued)

TABLE 9.1
(Continued)

Level	Classification	Summary description	Examples
		Radiation levels in an operating area more than 50 mSv/h. Exposure of a member of the public in excess of 10 mSv	**Penly Reactor, Seine-Maritime, France, 2012**: leak in the primary cooling circuit after a fire reactor 2 **Tricastin, Drôme, France, 2008**: environmental leak of 18,000 litres of water containing 75 kilograms of unenriched uranium
1	Anomaly	Overexposure of a member of the public in excess of the statutory annual limits Minor problems with safety components with significant defence measures remaining Low activity loss or stolen radioactive source, device or transport package	**Krško, Slovenia, 2008**: leakage from the primary cooling circuit **Atucha Reactor, Argentina, 2006**: reactor shutdown due to an increase of tritium in a reactor compartment **Tokaimura, Japan, 2006**: fire at the Japanese Atomic Energy Agency nuclear waste facility

away from the site looking after his family who were sick with flu. After receiving a call from his boss telling him to go to the site, he left his wife and two children telling them to keep all the doors and windows closed. Once on site, as soon as he understood the situation, he took off his personal radiation dosimeters to avoid having to leave work through exceeding his dose limit and climbed to the top of the 80 feet pile where he could look down through inspection holes in the concrete pile cap. Looking into the core of the reactor he could see the bright glow from the fire, as the workers below tried to disrupt the fuel cartridges with long steel rods. Tuohy decided that the extremely high temperature would melt the steel and was concerned that the concrete shield containing the core might collapse under the heat. Together with the site team, Tuohy had agreed that despite the risk of explosion, the worst-case solution was to extinguish the fire with water. Tuohy remined at his observation point for many hours whilst hoses were put in place to allow a gentle flow of water. When nothing happened, he switched off the air cooling fans and increased the water flow which eventually put the fire out. The fire lasted for 16 hours and caused approximately 740 TBq of iodine-131 to be released into the environment. A smoke filter constructed at the top of the chimney had successfully contained over 90% of the leaking radioactivity. Tuohy is regarded as one of the key individuals at Windscale who had averted a major disaster. Since he had removed his monitors his radiation dose was not known but he later left the nuclear industry and emigrated to Australia where he died at the age of 88. The Windscale fire was Britain's worst nuclear accident and at the time the government tried to minimise its consequences, releasing only brief details. However, to protect public health the sale of milk produced over an area of 500 square kilometres around the site was prohibited for several weeks following the accident because it was found to contain high levels of iodine-131. Studies of the workers who assisted in the operation have found no long-term health effects. Other studies have estimated that there could be between 100 and 200 deaths from cancer as a result of the incident. The Windscale reactor was subsequently sealed to contain the radioactivity and it was not until the late 1980s when the clean-up operation of the facility began. Arrangements for the safe disposal of nuclear waste at the Sellafield site are still ongoing.

9.3 THE SL-1 MELTDOWN, 1961

The SL-1, Stationary, Low Power Reactor Number One, was a US Army experimental reactor located at an army testing site 65 kilometres west of Idaho falls in Idaho. The prototype was constructed between July 1957 and July 1958 and went critical on 11 August 1958. The low power reactors were designed to replace diesel generators and boilers to provide electricity and heating for the Army's radar stations. On 21 December 1960 the reactor was shut down for maintenance, calibration and a refit of instruments. On 3 January 1961, three army specialist operators, John Byrnes, age 22, Richard McKinley, age 27, and Richard Legg, age 26, were on duty to restart the reactor. Byrnes was the reactor operator and he initiated the restart by manually withdrawing a control rod, but he misjudged the operation and took the rod out too far and much too quickly. As a result the reactor went critical in 4 milliseconds and the sudden generation of heat caused the reactor core to melt and vaporise. An exploding spray of water and steam knocked Byrnes to the floor killing him instantly. Legg was standing on top of the reactor when the shield plug on top of the vessel impaled him through his groin and shoulder, pinning him to the ceiling. McKinley, who was a trainee, was found nearby, unconscious with a head injury and died 2 hours later. This event is the only reactor accident in US history to have resulted in immediate fatalities. The only consoling aspect of this accident is that if the three men had not died from the trauma immediately following the explosion, they would have died a much slower death from acute radiation exposure. There were no other personnel on site, but after hearing the explosion and the alarms health physicists and firemen arrived on site, but they withdrew when their radiation monitors registered high radiation levels. The accident released about 3.0 TBq of iodine-131 and 41 TBq of other fission products into the atmosphere. The radioactive release was not considered significant due to the remote location of the reactor in the east Idaho desert. The bodies of the three casualties were buried in lead-lined caskets sealed with concrete and placed in metal vaults with a concrete cover. Some body parts were highly radioactive and buried in the Idaho desert as radioactive waste. The low power reactors were taken out of service and subsequent reactors were designed so that they could not be operated by withdrawing a single control rod. Numerous radiation surveys and clean-up operations have since taken place at the reactor site and surrounding area in the years since the SL-1 accident.

9.4 THREE MILE ISLAND ACCIDENT, 1979

The Three Mile Island nuclear power station was built near Harrisburg in Dauphin County Pennsylvania. The accident started on 4 am on Wednesday, 28 March 1979, in one of the two pressurised water reactors when a failure in the secondary cooling system was followed by a failure of the release valve in the primary cooling system which became stuck in the open position, allowing large amounts of coolant to escape. The emergency systems caused the turbine generator and the reactor itself to shutdown automatically, but the mechanical failures were compounded by misleading warning indicators in the control room and a lack of operator training. Radioactive gases from the reactor cooling system built up and a partial meltdown of the reactor core occurred with a small environmental release of radioactivity. The accident resulted in a loss of public confidence in the US nuclear power industry and led to the growing anti-nuclear movement. Because of the public concerns, the Pennsylvania Department of Health maintained a registry of more than 30,000 people who lived within 5 miles of Three Mile Island at the time of the accident for 18 years. The state's registry was discontinued in mid-1997, having shown no evidence of unusual health trends in the area. The average radiation dose to people living within 10 miles of the plant was 0.08 mSv with no more than 1 mSv to any single individual. The clean-up of the damaged nuclear reactor system at TMI-2 took nearly 12 years.

9.5 THE CHERNOBYL DISASTER, 1986

The Forsmark power plant situated north of Stockholm was Sweden's second largest nuclear power station. Cliff Robinson, a chemist who worked at the plant, described the events that took place during an interview for Radio Sweden in 2011. It was early morning on Monday, 28 April 1986, and Robinson had just finished his breakfast in the staff coffee room and went to the washroom to clean his teeth. On his way back to the locker room he had to pass through an entrance to one of the radiation zones where routine detector monitoring took place. As he passed the detector the alarm went off. This was unexpected since he had not been working in any of the controlled radiation areas. He went through three more times and on the final occasion the alarm did not sound. The monitoring staff first thought this was an error and promptly started checks to determine if the monitors required adjustment. Robinson carried on with his work but when he returned to pass the working zone monitor again, he saw a long queue of workers waiting at the detector point by the entrance. No one could pass through the control point because they kept on setting off the alarm. He borrowed a shoe from one of the workers and examined it with a germanium radiation detector to see if there was any evidence of radioactivity. The germanium detector could produce spectroscopic information by plotting the energies of any gamma rays that were detected. To his surprise the gamma spectrum grew at a much faster rate than he expected, showing that the shoe was highly contaminated with a wide range of different radionuclides. What was particularly alarming was that radionuclides detected were not those present in any of the cooling waters which could have leaked form the Forsmark reactor. His immediate thoughts were that there had been an explosion or a bomb which had dispersed radioactivity and caused contamination on the ground. Further alarms sounded, alerting all but essential staff to leave the plant. The radiation safety team carried out safety checks, repeating the results over and over again and within a couple of hours it became clear that Forsmark was not responsible for any release of radioactivity. The Swedish Radiation Authority were called in and further analysis showed that the radioactivity on the workers shoes was coming from the ground outside of the plant. By taking samples of grass and measuring the gamma spectrum they could identify the different radionuclides revealing that these were typical of the materials used in Soviet nuclear power plants.[1] Over the previous few days the wind had blown from the southeast and it had rained in the northern parts of Sweden, which had deposited radioactive fallout on the ground. The evidence pointed towards one of the nuclear power plants in the Soviet Union. At first the Soviet authorities denied that any release of radioactivity had occurred, but 2 days after the accident, the Soviet Union declared that there had been an incident at No 4 reactor of the Chernobyl nuclear power plant in northern Ukraine.

The Chernobyl power complex was situated about 130 km north of Kiev and about 20 km south of the border with Belarus. The plant consisted of four nuclear reactors of the Russian RBMK-1000 design. At 1:23 am on 26 April 1986 as No 4 reactor was being shut down for routine maintenance a test was being conducted to investigate reactor safety in the event of failure of the mains electricity supply to the plant. The test was to determine if in the event of a loss of power, the turbines could produce sufficient energy to keep the coolant pumps running until the emergency diesel generator was activated. In violation of the safety regulations the operators had switched off the control system that would prevent the reactor reaching unstable, low power conditions. Due to flaws in the RBMK reactor design a sudden power surge caused a steam explosion that ruptured the reactor vessel allowing steam to mix with the reactor fuel which caused a further explosion that lifted the 1,000 tonne cover plate off the top of the reactor, resulting in the largest accidental release of radioactivity into the environment in the history of nuclear power generation (Figure 9.1). With the control rods jammed the fuel became incandescent and an intense graphite fire continued to burn for 10 days releasing radioactive gases, condensed aerosols and a large amount of fuel and fission particles. To extinguish the fire initially around 200–300 tonnes of water per hour was pumped into the reactor but this was stopped to avoid flooding other reactor buildings. From the second to tenth day after the accident, around 5,000 tonnes of boron, dolomite, sand, clay, and lead were dropped on to the burning reactor core by helicopter to extinguish the fire and limit the release of radioactive particles.[2]

Two plant operators were killed as a result of the explosions. Other casualties included plant workers and fire fighters who climbed to the top of the turbine building to extinguish fires on the roof. Within 24 hours of the incident 28 workers including 6 firemen received what was estimated to be radiation doses of up to 20 Gy. All these workers developed symptoms of acute radiation syndrome including

---

[1] The process of identifying radionuclides by the investigation of their gamma spectra is known as nuclear forensics. Nuclear forensic science aims to identify the origin of nuclear materials found outside of regulatory controls and is used to combat theft, smuggling and illicit use of radioactive materials.

[2] The accident and the social and managerial culture of the time were depicted with widespread critical acclaim in the award winning Chernobyl historical drama, produced by HBO and Sky Television in 2019. The mini-series was exhaustively researched and the scenes were portrayed with a high level of accuracy; however, there were some factual discrepancies and dramatic inaccuracies.

FIGURE 9.1   Aerial view of reactor 4 at the Chernobyl nuclear plant in Chernobyl, Ukraine, showing damage to the roof of the reactor chamber following the explosion.

Source: Chernobyl reactor. www.shutterstock.com/editorial/image-editorial/ukraine-chernobyl-30-years-later-chernobyl-ukraine-7382120a

nausea, vomiting, diarrhoea, headaches, burns and fever and all died by the end of July. Underlying bone marrow failure was the main contributor to all deaths that occurred during the first 2 months. Another primary cause of death was considered to be due to infection from extensive skin burns from beta radiation. Around 1,000 workers were brought on site in the first few days, many of whom received high radiation doses. The operation continued to secure the site and shield the damaged reactor, so that the remaining three reactors could be restarted. Around 200,000 people, 'liquidators', were recruited from all over the Soviet Union between 1986 and 1987. They received high doses of radiation, averaging around 100 mSv. Some 20,000 liquidators received about 250 mSv, with a few receiving as high as 500 mSv. In all around 600,000 people were involved on site, but most of these received only low radiation doses more comparable with the background levels of around 3 mSv per year.

The total release of radioactive substances was considered to be greater than 14 exabecquerels (EBq, 14×10^{18} Bq), which included 1.8 EBq of iodine-131, 0.085 EBq of caesium-137 and other caesium radionuclides, 0.01EBq of strontium-90 and 0.003 EBq of plutonium radionuclides. The dispersal of iodine-131 presented an immediate problem to the local population. All 45,000 residents of the power plant town of Pripyat were evacuated on April 27. By May 14, around 116,000 people who had been living within a 30-kilometre radius had been evacuated and relocated. About 1,000 of these residents returned unofficially to live in the contaminated zone preferring to live in their own homes despite the health risks. Most of those evacuated individuals received radiation doses of less than 50 mSv, although a few received 100 mSv or more. Large areas of Europe were affected and an area of more than 200,000 square kilometres was contaminated with caesium-137. The three most affected countries were Belarus, the Russian Federation and Ukraine, but the contamination affected countries throughout Europe. After the initial period, caesium-137 became the nuclide of greatest radiological importance. Ground deposition was strongly influenced by the prevailing winds and rainfall, but the end result was that sheep in highland areas as far away as Wales had consumed grass contaminated with caesium-137 and farmers were prohibited from selling their lamb at food markets for many weeks after the incident.

Despite the large amount of radioactivity released as a result of the accident the health impact does not appear to be as great as might have been expected. There have been some exaggerated figures of the death toll attributable to the Chernobyl disaster; however, these have been discounted. In February 2003, the IAEA established the Chernobyl Forum, in cooperation with seven other United Nations organisations and the competent authorities of Belarus, the Russian Federation, and Ukraine. The conclusions of the Chernobyl Forum study were that apart from an increase in the incidence of thyroid cancer, there was no evidence of a major public health impact attributable to radiation exposure 14 years after the accident. The United Nations Scientific Committee on the Effects of Atomic Radiation has concluded that there were some 6,500 cases of thyroid cancer; however, there are very effective treatments for thyroid cancer and only 15 fatalities have been recorded. There was little evidence of any increase in radiation-induced leukaemia which has a latency period of 5–7 years, even among clean-up workers where it might be most expected. Overall, as of 2005, fewer than 50 deaths have been directly attributed to radiation exposure from the disaster. Mental health coupled with smoking and alcohol abuse has been identified as a far greater problem than radiation, mainly because the accident happened at the time when the population of the region had relatively poor underlying levels of health and nutrition.

9.6 THE TOKAIMURA ACCIDENT, 1999:
A FATE WORSE THAN DEATH

The Tōkai Nuclear Power Plant is located in the village Tokaimura in the Ibaraki Prefecture of Japan about 130 km northeast of Tokyo and facing the Pacific Ocean. The village has a population of 34,000 and has many facilities concerned with the nuclear industry such as The Japan Atomic Energy Research Institute and The Japan Nuclear Cycle Development Institute. The power plant was built to the British Magnox design to generate power and was decommissioned in 1998. A second reactor was built in the 1970s but has not been in operation since it shut down automatically in 2011 due to the Tōhoku earthquake and tsunami. Two nuclear accidents have occurred at the plant: one due to a small explosion in 1997 and one due to criticality accident in 1999. The second accident was the most serious and led to what is probably the most gruesome account of the medical care of acute radiation sickness.

On the morning of 30 September 1999, 35-year-old Hisashi Ouchi said goodbye to his wife and son and left for work at the Tokaimura nuclear fuel processing facility. Ouchi was one of three technicians who had been working on a procedure to dissolve and mix high-purity enriched uranium oxide with nitric acid to produce uranyl nitrate fuel for the experimental Joyo fast reactor. This was the first time he had worked in the conversion test building and together with his colleagues, Masato Shinohara and Yutaka Yokokawa, they started the last few steps for preparing the reactor fuel. The standard operating procedure was for the uranyl nitrate to be stored inside a dissolution tank and then to be slowly pumped into a water-cooled precipitation tank, but the work was behind schedule and so the plant manager had approved a quicker and more convenient method using stainless steel buckets. The technicians had not been trained for the work they were doing and they were not aware of the dangers involved in the work. Following his boss's instructions, Ouchi first filtered the uranium-235 solution that had been melted in the bucket and his boss and colleague poured the solution into the precipitation tank. His boss placed a funnel in a hole at the top of the tank and held it in place whilst his colleague stood at the top of a ladder and poured the solution into the tank. After Ouchi had finished the filtration step, he took over from his boss standing next to the tank to hold the funnel. At around 10:30 as they poured the contents of the seventh bucket, there was a bright blue flash of Cherenkov radiation and the sound of a loud crack immediately followed by the sound of radiation alarms. The constraints of the non-approved working method had changed the concentration and volume of the mixture in the containers. This had started a critical chain reaction that continued over several hours. Ouchi and Shinohara immediately felt pain, nausea and had difficulty in breathing. They ran out of the work area and Ouchi went to the changing area. He vomited and briefly lost consciousness. The three workers were taken to the nearby National Mito Hospital and then transferred by helicopter to the National Institute for Radiological Sciences. In the Tokaimura processing building the chain reaction continued overnight and was finally stopped the next morning when a volunteer group of workers known as the 'suicide squad' drained water from the cooling jacket around the tank and added boron to the liquid. The cooling water had been acting as a neutron reflector directing the neutrons into the tank, accelerating the critical reaction. Twenty-four workers received doses of up to 48 mSv, these being too small to cause acute radiation effects. In total 436 workers received elevated radiation doses as a result of the accident.

Radiological tests carried out at the Radiological Sciences Institute by the health physics team 5 hours after the event showed some gamma and beta radiation but no evidence of alpha radiation. Ouchi had been sick in the ambulance and his vomit had been collected in plastic bags. Analysis of the vomit showed the presence of activated sodium-24, confirming that a criticality event had occurred and neutrons had been emitted. Ouchi had received the highest effective radiation dose of between 16 and 20 Sv. Shinohara's dose was between 6 and 10 Sv and the boss, Yokokawa, received between 1 and 4.5 Sv. Of the two technicians that received critical doses from the incident, it was Hisashi Ouchi who became the subject of greatest interest. Ouchi was being treated by Kazuhiko Maekawa, a Department of Medicine Professor at the University of Tokyo. Professor Maekawa was a 'hands on' physician who had only recently become chairman of the Nuclear Safety Research Association's Radiation Emergency Treatment Task Force and had quickly become aware of the inadequate education for physicians who may have to work in emergency radiation medicine. When Prof Maekawa first met Hisashi Ouchi over 24 hours after the incident, he could not believe his eyes. Ouchi had been a rugby player in his youth. His height was 174 centimetres and he weighed 74 kilograms. His face was a bit swollen and his eyes were a little bloodshot, but his skin was a healthy looking red colour with no sign of peeling and there were no blisters. When asked if he was in pain Ouchi said his right hand was painful and he had a pain under his ear. Maekawa later commented that despite Mr Ouchi having received the highest radiation dose of the three men he seemed the most emotionally stable and despite his declining white blood cell counts he thought they might be able to save his life. He felt pity for his patient and wanted to give him the best that modern medicine could provide. To the concern of his junior colleagues, he was heard to say that letting a patient die was a dishonourable act for a physician.

Maekawa knew that if he was to treat the first criticality patient in Japan, he needed cooperation of specialists in haematology, gastroenterology and dermatology and he needed his patient in an intensive care unit. Hisashi Ouchi expressed his thanks after being transferred to a private room in the intensive care unit at the University of Tokyo Hospital. Initially the intensive care nursing staff were afraid of exposure from secondary radiation, but their fears were dispelled when a radiation monitor from the radiology department demonstrated that there was nothing hazardous to detect. The doctors wrote up his care plan and his treatment started. On day 6 after the accident, bone marrow cells that had been taken the day after Ouchi had been admitted to the hospital were examined under the microscope. Analysis of the chromosomes showed that instead of the normal 23 clearly identified pairs, only fragments and fused agglomerations could be seen. The doctors were concerned that his white blood cell and platelet counts were dropping and were now considering haematopoietic stem cell transplantation using cells from a healthy donor, a procedure often used to treat leukaemia patients. They carried out computer searches of suitable donors and eventually found that Ouchi's younger sister was the most suitable donor. When she was approached, his sister begged the team to take as much of her blood as they needed if it would save her brother. On the morning of day 7 after the accident, Ouchi's sister attended the blood transfusion centre on the third floor of the University of Tokyo Hospital central building and lay on a bed for 4.5 hours connected to a blood extraction filter system that removed 160 millilitres of haematopoietic stem cells from her blood and returned the remaining blood components to her circulation. By 3 pm in the afternoon Ouchi received his sister's blood cell. The entire procedure was repeated the next day. The doctors then had to wait for the following 10 days to determine if the cells were taking root and helping to grow new blood cells.

As the days passed the medical team contributing to Ouchi's care grew to a total of 13 separate departments. His condition was gradually changing and he underwent a series of regular invasive investigations. Blood and bone marrow samples were taken for assessing his blood count, tissue samples removed from his nose, throat and skin to test for infection and X-ray and CT examinations were carried out to examine his internal organs. Ouchi said that the procedures were tiring but that he should not complain. His thirst was increasing and his breathing became laboured. The X-rays had shown a mass in the right lung, the side of his chest that had been nearer the tank at the time of the accident. It was decided to carry out a pleural puncture to drain the liquid from his chest. At this point Ouchi made numerous outcries that he could not take it anymore and pleaded for them to stop, shouting that he wanted to go home and did not want to go on with the treatment. The nursing staff were aware that he was suffering and the pills Ouchi had been given to help him sleep were replaced with more powerful sedatives. Throughout his treatment, Ouchi's family had been visiting him every day. They had developed a close relationship with Prof Maekawa and believed that the doctors had his best interests at heart.

By day 10 Ouchi's skin started to peel off. This was first observed when removing tape that was holding intravenous lines and electrocardiogram electrodes, but this became worse each time the nurses washed and dried him. Around this time a Professor of Medicine from University of California's Haematology-Oncology Division, Robert Gaye, arrived in Tokyo. Professor Gaye had previously performed haematopoietic stem cell transplants on 19 victims of the Chernobyl Disaster. Over 17 days Gaye worked with the team attending meetings and advising on the treatments. On day 11 Ouchi's condition deteriorated and a tracheal tube was inserted to provide an airway for breathing. The team continued to monitor his blood counts to see if the stem cell transplantation had produced any beneficial effect. By day 17 his white cell count which had been continually dropping started to increase. A further bone marrow biopsy was taken and by examining the chromosomes they could identify the female sex chromosome from his siters cells confirming they had taken root and were producing new blood cells. The red cell, white cell and platelet counts in Ouchi's blood continued to rise. Prof Maekawa was feeling optimistic.

Ouchi's treatment continued but he was still in a desperate critical condition. Because of the damage to the lining of his throat he could no longer take food orally and was fed intravenously. Despite the fragile nature of his intestines an endoscopy was carried out using a thin tube to minimise any further damage. This was a particularly stressful and difficult procedure for the doctors to undertake since with this scoping procedure there is always a risk of perforating the bowel which would require immediate surgery, a procedure that Ouchi would not have survived. The conclusion from the endoscopy proved to be better than expected and the lining of the gut appeared to be intact. He continued to be treated with various drugs, including some that were undergoing clinical trials in Japan at that time.

Ouchi's case became of major interest to radiation specialists around the world. On October 28, experts arrived from the US, France, Russia and Germany. They had never witnessed anyone who had survived for so long after receiving such a high dose of radiation. As the days passed, skin loss became a major issue. The skin on Ouchi's swollen right hand was the first to disappear. His whole right arm and chest wall became covered with open blisters and the skin loss then extended to most of the front side of his body, including his feet. Eventually it took 10 people to change his dressings each day. This was carried out by applying antibiotic ointment and covering his entire body with smooth medical gauze. His fingers were individually wrapped to prevent them from sticking together and his eyelids would no longer close. By day 50, the medical team considered growing cultured skin from his own skin cells but the chromosomes in his skin cells were so badly damaged that the cells would not grow normally. Eventually they used a skin donation taken from his sister's thigh but Ouchi's continuing skin loss was becoming a major issue, as he was losing large amounts of fluid through his body surface. By weighing him daily, they could determine that he was losing over 2 litres of fluid a day. Ouchi continued to lose greater amounts of fluid through internal bleeding from his stomach which was producing large amounts of bloody diarrhoea. This amounted to a loss of over 10 litres of fluid a day and as a result it was proving difficult to maintain his blood pressure. The drug treatments and personal care was taking a huge amount of hospital resources. Most members of the medical and nursing teams were questioning if this should go on, yet in the clinical meetings Prof Maekawa continued to consider new treatment plans.

By 2 months his body was entirely covered in gauze and his limbs were suspended from wire frames to ease any sores. At around 7 am on Saturday, November 29, day 59, the hospital resident doctor was nearing the end of his night shift and doing some observations at Ouchi's bedside. Ouchi's airway respirator and ECG electrodes had been briefly removed so that a chest X-ray could be taken. When the breathing tube and electrodes were reconnected, he was no longer breathing and his heart was not beating. After enduring a massive dose of radiation and 59 days of intensive and painful investigation Ouchi had suffered a cardiac arrest.

At that point Maekawa came in shouting 'cardiac arrest get some doctors in and give him Epinephrine'. After giving the strong cardiotonic drug and applying cardiac massage his heart started beating. By 7.25 it had stopped again. It had stopped for a total period of nearly 50 minutes but after further injections and CPR his heart was beating once more and he was breathing spontaneously. There were serious concerns that many of Ouchi's organs had been damaged, including his brain, liver and kidneys. Many of the junior medical staff and nurses were confused and depressed but Ouchi's family still believed in him and were giving Maekawa their support.

Prior to the cardiac arrest Ouchi had been able to communicate to his family with facial expressions. Now, he no longer responded to anyone and his brain no longer responded to any form of stimulus. He was kept alive by artificial breathing, plasma exchange and by giving a plethora of drugs for a further 22 days after his cardiac arrest. Finally at 11:21 pm on December 21 Hisashi Ouchi's heart stopped, after a slow and painful death that lasted 83 days. His heart was probably the very last organ in is body to cease functioning, all the other organs and tissues had succumbed to the effects of acute radiation poisoning.

Masato Shinohara, who received a lower radiation dose, was also treated at the same hospital and endured radical cancer treatments, numerous successful skin grafts and blood transfusions to boost his stem cell count. After a 7-month battle, he was unable to fight radiation-induced infections and internal bleeding. He died of multiple organ failure at the age of 40 on 27 April 2000. The technical supervisor Yutaka Yokokawa, who was 54 years, received treatment from the National Institute of Radiological Sciences in Chiba. He had minor symptoms of radiation sickness and was discharged 3 months later. In October 2000 he faced negligence charges along with five other employees of the plant.

9.7 THE FUKUSHIMA DAIICHI DISASTER, 2011

On 11 March 2011, shockwaves from the Tōhoku earthquake caused the reactors at the Fukushima plant to shut down their normal power generating operation. Electrical supply problems caused the emergency diesel generators to take over supplying power to maintain flow of the coolant through the four reactor cores. The emergency systems worked well until a 14-metre high tsunami also caused by the earthquake breached the plant's sea wall, flooding the lower reactor chambers and stopping the emergency generators from working. Three reactors went into nuclear meltdown causing three explosions, releasing over 27 PBq of caesium-137 into the Pacific Ocean. Other environmental emission of iodine-131, caesium-134 and cea-sium-137 also occurred. Concerns of larger radiation releases led to the setting of a 20 kilometre exclusion zone around the site and around 154,000 residents were evacuated. Widespread contamination of water food and fish was later observed. There were no cases of acute radiation exposure and no deaths resulting directly from radiation poisoning. The first acknowledged death from radiation at the site was confirmed in 2016 when a plant worker in his 50s died from lung cancer. It was found that six workers received radiation doses in excess of 250 mSv and 167 workers received doses above 100 mSv. Around 18,500 people died from the effects of the earthquake and tsunami. Monitoring of the activity in sea water demonstrated widespread contamination in both the northern and southern oceans and measurable radioactivity reached the coast of Russia on the March 14 and the US coast 2 days later. Since the event, scientists have discovered increased traces of caesium-137 in wine grown in a vineyard in the Napa Valley, California. Restoration of the Fukushima site is still ongoing and it is understood that amount of contaminated water continues to build up reaching the limit of the site holding capacity by 2022. Environmental organisations such as Greenpeace are concerned that further discharges of into the ocean will be harmful to human health.

9.8 MILITARY AND MEDICAL MISHAPS

In addition to the accidents at nuclear power plants there have been many other incidents with release of radiation resulting in poisoning. In fact, there have been far too many to describe in detail here but the most serious military accidents included those taking place in nuclear-powered submarines and on aircraft carrying nuclear weapons. One of the most notable accident was the K-19 submarine which was the first of two Russian vessels to carry nuclear ballistic missiles. In July 1961 the submarine was on exercises in the North Atlantic when the reactor coolant system developed a leak causing the pumps to fail, and although the reactor did not explode the temperature rose to dangerously high levels at around 800°C. As a result, radioactive steam containing fission products was drawn into the ship's ventilation system and radioactivity spread to the other compartments of the submarine that were occupied by the crew. Captain Nikolai Vladimirovich Zateyev was unable to contact Moscow for help and sent seven engineering crew members to the high radiation area to fabricate a new coolant system by cutting off an air vent valve and welding a water-supply pipe to the cooling system. All seven members of the engineering crew and their divisional officer died of radiation exposure within the following month. Fifteen more sailors died within the next 2 years. The K-19 incident and fate of the sailors was kept a secret until after the breakup of the Soviet Union, when the newspaper *Pravda* revealed that radiation had killed many members of the crew. The accident was the subject of the 2002 movie *K-19: The Widowmaker*.

There have been a number of radiological incidents including miscalculation, incorrect calibrations, equipment malfunctions and human errors in radiotherapy treatments, resulting in treatment overdoses which have led to tragic consequences for the patients and the staff treating them. There have also been some serious incidents involving medical radioactive treatment sources, such as those in Goiania in Brazil, Mexico City, Thailand and Mayapuri in India. The Goiania incident is regarded as one of the most serious radiological accidents to have occurred and led to widespread consequences across an entire city. Towards the end of 1985 a private radiotherapy institute, the Institute Goiano de Radioterapia, moved to new premises. As part of the move, they had to relocate most of the equipment that was used for diagnosis and treatment, including a cobalt-60 teletherapy unit for treating patients with external radiotherapy. The hospital left one old teletherapy unit that contained a sealed 50.9 TBq source of caesium-137, since it was not required at the new site. After the old building was vacated, no one notified the licensing authority as was required under the terms of the institute's licence for holding radioactive sources. These teletherapy units were designed so that the radioactive source was held in a sealed container, shielded on all sides by very thick lead. For delivering treatment, the patient was placed on a couch and the head of the unit containing the source was moved mechanically around the patient in the same way a linear accelerator treatment unit is used today. The source was contained in a rotating wheel that could be mechanically rotated to a small opening in the shielding allowing a fine focussed beam of gamma rays to be directed at the treatment site. Once the prescribed dose has been given, the wheel automatically turned back to its storage position blocking any radiation from reaching the patients or staff. The unit was abandoned with the source in the closed position and would not have represented an immediate danger to anyone who entered the treatment room. On 13 September 1987 Roberto dos Santos Alves and Wagner Mota Pereira illegally entered the partially demolished building in search of anything that might have some scrap value. They found the teletherapy unit and dismantled as much of it as they could, placing the source assembly in a wheelbarrow and taking it back to Alves's home. During the evening, as they were taking the wheel mechanism apart they both began to vomit due to radiation sickness. The next day, Pereira suffered from diarrhoea and dizziness, and his left hand became swollen and burnt (he eventually needed surgery to amputate several fingers). On September 15, Pereira visited a local clinic where he was diagnosed with food poisoning and he was told to return home and rest.

Alves continued to dismantle the equipment and removed the small capsule from its protective rotating head. The capsule contained around 93 grams of caesium chloride powder. His prolonged exposure to this material eventually led to ulceration of his right forearm which later required amputation. On September 16, Alves succeeded in puncturing the capsule with a screwdriver, allowing him to see a deep blue light coming from the small opening he had made. At first he thought this was kind of gun powder and tried to light it, but the powder did not ignite. Two days later Alves sold the items to a nearby scrapyard. That evening, the owner of the scrapyard, Devair Ferreira, noticed the blue glow from the hole in the capsule and thought it was valuable or perhaps even supernatural, so he took it into his house to show his friends and family the strange glowing substance. On September 21, one of Ferreira's friends freed several rice-sized grains of the glowing material from the capsule and Ferreira handed them around to his friends and family members. The same day, his 37-year-old wife Gabriela Ferreira began to fall ill. On September 24, Devair's brother manged to scrape some dust out of the source and took it to his house a short distance away, where he spread some of it on the concrete floor to watch the blue glowing particles. His 6-year-old daughter, Leide das Neves Ferreira, was amused by this and scattered it over her body to show it off to her mother. Dust from the powder fell on a sandwich she was consuming and she eventually absorbed 1 GBq resulting in a dose of 6 Gy. Gabriela Ferreira was the first to notice that many people around her had become severely ill at the same time. On 28 September 1987, 15 days after the source was stolen, she took some of the material from the scrapyard in a plastic bag to a hospital. Fortunately the plastic bag reduced the level of contamination at the hospital. On the morning of September 29, a visiting medical physicist used a scintillation detector to confirm the presence of radioactivity and persuaded the authorities to take immediate action. By the end of the day, the city, state and national government became only too well aware that a major radiological incident had occurred. Four of the victims ultimately died and 28 people suffered radiation burns. A huge monitoring operation was carried out and around 112,000 people were examined for radioactive contamination, and 249 of them were found to have been contaminated. Large areas of contamination were found and contaminated money had to be taken out of circulation. The decontamination operation required the demolition of seven residences and the removal of large areas of topsoil. This produced large quantities of radioactive waste and a temporary waste storage site was chosen 20 km away from Goiania. Various types of packaging containment were used, according to the different levels of contamination. The removal and storage of waste required 3,800, 200 litre metal drums, 1,400 metal boxes, 10 shipping containers and 6 units of concrete waste containment.

Whilst this chapter has described only some of the many nuclear and radiological incidents that have taken place, it does provide a picture of the kind of situations that have led to major accidents. These incidents have been the result of neglect, carelessness, cost cutting, ignorance or criminal acts and some have simply been most unfortunate accidents. The consequences of these events have affected many people throughout the world. Some people faced nothing more than a temporary inconvenience but others faced unexpected pain and personal tragedy.

10 ☢

Home and Away

Thank God, if it was going to be anybody living in that house it would be me,
somebody who through their work could discover the situation.
Comment made by Stanley Watras when interviewed in 1985.

10.1 RADON ALERT

In 1984 Stanley Watras and his young family moved into their stylish suburban home
in Boyerstown, Pennsylvania. Stanley was a 34-year-old construction engineer who
had taken a building job at the Limerick Nuclear Power Plant which was commis-
sioned by the Philadelphia Electric Company in Pottstown, next to the Schuylkill
River in Montgomery County. Before any nuclear fuel had been brought on site, the
plant was being fitted out with the internal furnishings, electrical wiring and safety
equipment. As part of the standard safety equipment, radiation monitors were fitted
at the exit doors and left on for a period of background radiation testing prior to the
arrival of any radioactivity on site. As soon as the plant was operational it would be
mandatory for the staff to monitor themselves for the presence of any radioactivity
on their skin or clothing whenever they left the building. This would then provide
reassurance that there were no environmental leaks of radioactivity and that the staff
were following the correct safety procedures.

The morning of 2 December started as an ordinary day for Stanley as he headed
to work. After arriving on site, he entered the building and suddenly there was the
sound of alarms, buzzers and flashing red lights. Stanley looked around puzzled as
the plant staff went to check the radiation detectors around the doors assuming that
a fault had developed. They checked the alarms and then checked Stanley. There
was no radioactivity on site, so they decided it must have been a spurious event.
When Stanley walked into work the next day the staff could not believe it when the
same thing happened. On checking him with radiation monitors, the health physics staff team were surprised to find that he was contaminated with radioactivity.
When this occurs, the standard radiological practice was, and still is, to remove any
contaminated clothing and wash the skin to remove any radioactivity. In Stanley's
case this had to be done before he started his working day. The dilemma was that
there was no radioactivity on site, so it was difficult to understand how Stanley was
becoming contaminated. Even more puzzling was that when Stanley went home at
the end of the day there was little evidence of any further radioactive contamination.
Unfortunately, Stanley continued to set off the radiation alarms as he entered the
building every morning for the following 2 weeks. Each time he was monitored, the

staff found that his clothes were a major source of the contamination and so he was put through the decontamination procedure each day before he started work. The radiation safety section decided they needed to look elsewhere for possible sources of contamination, so a team of specialists went to Watras's home to carry out a radiation survey. They were astonished when they found that the house contained high levels of radioactivity. This was all around in the air, circulating between the rooms and contaminating household items. As they went down to the basement, the survey monitors showed increasing radiation levels that were as high as 100,000 Bq per cubic metre of air. This was the highest level of radiation that had ever been recorded in an American home and was around 700 times higher than the accepted safe level for human exposure. After further investigation it was found that the Watras family home and other nearby houses were built over a 10 metre wide geological fracture that ran through an underground vein of natural uranium. The source of the radiation turned out to be radioactive radon gas, a decay product of uranium that was escaping from the rocks beneath and rising up into the house. It was evident that the entire family was living in an environment similar to that of a uranium mine, and they had been at a similar health risk to smoking 135 packets of cigarettes per day. Naturally, Stanley and his wife Diane were anxious about the health effects on themselves and particularly on their children. The family immediately moved out of their house and the US Environmental Protection Agency and Pennsylvania health officials moved in, turning their home into a laboratory for the long-term measurement of radon and radon decay products. It took many months before the technicians were able to reduce the radon concentration to an acceptable level by fitting an impermeable membrane in the basement and installing vents and air-flow systems. They also had to install a radon detection system with an alarm that would trigger if the radiation level increased above the safety limit. After the mitigation work and further testing the family was able to return home.

Radon is a gas found in Group 0 of the periodic table. It is colourless and has no tase or smell. It is a natural decay product of radium-226 which occurs in the chain of radioactive decay from uranium-238 through to stable lead-206. Uranium is commonly found in the earth's crust but is most concentrated in sedimentary and volcanic rock formations that can be found at various locations around the world.

The highest levels of radon have been found in Scandinavia, the US, Iran and the Czech Republic. Being a gas, radon seeps out of the earth and into the atmosphere. Radon-222 with a half-life of 3.8 days is the most stable radionuclide of radon and is responsible for the majority of public exposure to background radiation. It becomes a particular hazard to health when it enters buildings and closed spaces with poor air flow. The increasing use of building insulation and double glazing has resulted in the build-up of dangerously high concentrations of radon in some domestic properties, schools and workplace buildings. The main health hazard from radon is from lung intake through normal breathing. The biological damage occurs from the radon decay products, mainly polonium-218 and polonium-214 which become deposited in the bronchial tissues of the airways delivering the majority of the radiation dose to the lungs in the form of alpha particles. Epidemiological studies have shown conclusively that this is carcinogenic to humans and can cause lung cancer. In fact, radon is considered to be the second largest cause of lung cancer after cigarette smoking.

Low levels of radon can also be found in drinking water but the release of the gas from water is negligible compared to other sources. Scientific studies have shown that the risk of stomach cancer and other gastrointestinal malignancies from radon in drinking water is small.

The fact that Stanley Watras was working at a nuclear power plant turned out to be particularly fortunate. Not surprisingly, the incident received a great deal of publicity in the US. Radon was subsequently found in every region and it was shown that around 6% of US houses had elevated levels. In 1988 the US Congress responded to national concerns and passed the Radon Abatement Act. This act called for the reduction of indoor levels of radon down to outdoor radon levels. In the same year, the US Environmental Protection Agency declared that 'contamination of homes across the nation by cancer causing radon was the nation's most serious air pollution problem' and recommended that virtually every home in the USA be tested for radon. Similar testing has been carried out in many other countries around the world.

In the UK the National Radiological Protection Board (NRPB) was formed in 1970 by the Radiological Protection Act. The remit of the board was to conduct research, give advice and provide technical services in the field of radiation protection.[1] In the early 1970s the NRPB considered that indoor exposure to radon was likely to be the most significant source of radiation exposure to the general public in the UK. A number of surveys were undertaken and advice to the public was first issued in 1987. The surveys allowed high risk areas to be designated. These areas included the counties of Devon, Cornwall, Derbyshire, Northamptonshire and Somerset. Further regions in Scotland and Northern Ireland were identified in 1993. In 1996, regions of Wales were also declared as affected areas. Public reaction to these findings was mixed. The NRPB sent questionnaires to around 10,000 homes in Cornwall and Devon which had some of the highest high radon levels. Around 50% of the households returned their completed questionnaires and only 10% of them had carried out any form of remediation. The cost of remediation was cited as being the biggest barrier. Around 53% said that it was too expensive whilst 28% stated that it was not important for people of their age and 27% stated that they were not convinced that radon was a serious risk to health. National environmental services and the building industry has an increased awareness of the issue in new buildings but there is clearly a need for further education about the potential harm from radon for those living in older buildings. In most countries owners can now obtain radon gas detectors and install radon mitigation solutions, such as an impermeable radon floor membrane, underfloor gas sumps, positive pressure air systems and fan-assisted ventilation.

10.2 HOME TRUTHS

Radon gas detectors are now widely available for domestic use. The use of these monitors has led to some unexpected findings. An article in the journal *Epidemiological Health* published in 2019 gave an account of a homeowner in Korea who was using

---

[1] At the time of writing Public Health England is the UK's primary authority on radiation, chemicals and environmental hazards. They provide a complete range of radiation protection services including radiological assessment, personal dosimetry and radon measurement services.

a radon detector, when he unintentionally discovered that the radon and thoron concentrations in the mattress of his bed exceeded the national indoor radiation limit. Since people spend a significant part of the day (or night) in bed sleeping, there was a great deal of public interest. This led to concerns that the internal radiation exposure from radon and thoron to users of this type of mattress was a serious social issue. The mattress in question contained monazite, which is a phosphate mineral and a source of radon and thoron. The thoron concentration from the monazite was found to be 10 times higher than that of radon. Many similar mattresses from the same manufacturer were found to emit radiation resulting in doses higher than the annual dose limit of 1 mSv designated by the environment protection guidelines in South Korea. The natural background radiation exposure among the general population of South Korea is around 3 mSv per year, half of which is caused by internal exposure to radon.

Excluding radon and possibly some old family heirlooms in the attic, such as watches, clocks, Vaseline glass and medical appliances that may have contained radioactivity, the only other source of radioactivity you might find in the home might be smoke detectors. These life-saving devices often contain a small amount of americium-24, amounting to around 37 kBq. Americium-241 is an alpha emitter with a

FIGURE 10.1   Location of the americium-241 source in a domestic smoke detector.

Source: Location of the americium-241 source. www.shutterstock.com/image-photo/old-domestic-smoke-detector-alarms-on-1100184704

half-life of 432 years. As it slowly decays, it produces a negligible amount of gamma radiation, so there is no external hazard to people in rooms and building. These smoke detectors have two ionisation detectors: one is sealed and works as a reference chamber and the other is open to the air. Under normal conditions the current in both chambers is the same, but if smoke particles enter the open chamber a change in current is detected and this activates the alarm. Some European countries and American states have now prohibited the use of this type of alarm in favour of the newer optical photoelectric detectors that do not contain any radioactivity. A potential hazard exists with radioactive smoke detectors if a large number of devices are not stored securely or if they are disposed of inappropriately. The americium-241 in these devices is contained in a sealed container and is only a risk to health if the container is punctured and the material extracted. A person would have to open the sealed chamber and ingest or inhale the americium for the risk to be significant and why would anyone want to do that?

10.3 THE RADIOACTIVE BOY SCOUT

On 13 August 1994, an 18-year-old teenager was stopped by police whilst driving through Clinton township in Michigan. The police were responding to a call concerning a young man who had allegedly been stealing car tyres. They opened the back of the car looking for tyres, when they came across something quite different. A red toolbox sealed with duct tape and an assortment of objects including 50 foil-wrapped cubes containing grey powder, small disks, cylindrical metal objects, fireworks, a clock face, gas lantern mantles, vacuum tubes and various chemicals and acids. The young driver warned the police that the grey powder was radioactive and they immediately thought this was some kind of nuclear device or a dirty bomb. They towed the car to the police station (something that was perhaps considered to be unwise after the event) and proceeded to look into the matter. The teenager was David Hahn, who turned out to be a rather troubled and obsessive young man. His parents were separated and he had become increasingly withdrawn and secretive. In his younger days he was often seen wandering the streets, his pockets bulging with copper wire, electrical connectors, transistors and a soldering iron. This had given him a reputation as Clinton township's young mad scientist. After reading *The Golden Book of Chemistry Experiments* he became obsessed with chemistry and spent hours and days conducting home chemistry experiments, which would cause small explosions and other mishaps in his bedroom. His main outlet seemed to be taking part in Boy Scout activities and he worked hard to earn a Boy Scout Merit Badge in Atomic Energy. Such an award was not that unusual, since many school and college students carry out experiments with small sources of radioactivity but they would not be allowed to take the materials home with them. In the 1950s, the Gilbert Toy Company in America produced chemistry sets, metal construction sets and also the 'Atomic Energy lab'. The lab came in a smart presentation box with jars containing small amounts of uranium-238, lead-210 and zinc-65, a cloud chamber and a battery-powered Geiger counter (Figure 10.2). The instruction book described awe-inspiring experiments, for watching the path of electrons and alpha particles and a 'hide and seek kit' which challenged players to locate radioactive samples hidden around the house.

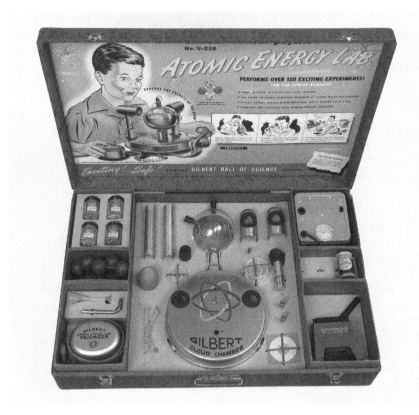

FIGURE 10.2   Photograph of the Gilbert Atomic Energy Lab which was sold in the US in the 1950s for $49.50.

Source: Photograph of the Gilbert Atomic Energy Lab. Printed with permission from the Oak Ridge Associated Universities Health Physics Historical Instrumentation Collection www.orau.org/ptp/museumdirectory.htm

David Hahn was incentivised by his Boy Scout Merit Badge and decided that he would try something a bit more ambitious; he would build a nuclear reactor in his mother's garden shed. Armed with a Geiger counter, over the next few years he collected a large amount of household products that contained radioactivity, including radium from clock dials, tritium from gunsights, thorium from gas camping mantles and americium from smoke detectors. By stealing smoke detectors and buying up stocks of damaged goods he accumulated a large amount of americium-241. He even posed as a college teacher and wrote letters to obtain radioactive sources that were gladly sent to him without question. When he was apprehended by the police in 1994, he was in the process of dismantling his experiment and had thrown much of the activity away with the household rubbish. He never obtained sufficient material to produce a critical reaction and it is thought that he ended up trying to produce a neutron source. Once the details emerged, the police initiated a Federal Emergency involving the FBI and the Nuclear Regulatory Commission and his mother's property was designated a hazard zone. To her neighbour's concern, men wearing radiation

FIGURE 10.3   Building 7, Gvardeytsiv Kantemirovtsiv Street, Kramstorsk, Ukraine.

Source: Building 7, Gvardeytsiv Kantemirovtsiv Street, Kramstorsk Public domain photograph. Photograph in the public domain from Artemka, under the creative commons Attribution-Share Alike 4.0 International licence

suits cordoned off his mother's garden for decontamination and removal of the materials which were disposed in a low-level radioactive waste repository. It was considered that the amount of radioactivity amounted to over 1,000 times background levels. Hahn never agreed to a medical examination, but the health protection scientists believed that his life expectancy might have been greatly shortened, because he had spent long periods in a small, enclosed shed with large amounts of radioactive material using minimal safety precautions. Hahn became depressed after the incident. He broke up with his girlfriend and his mother committed suicide in 1996. He was later charged with removing a number of fire detectors from an apartment building where he was living. After a short career in the military, at the age of 39, Hahn died due to intoxication from excessive misuse of alcohol and drugs.[2]

10.4 THE KRAMSTORSK TRAGEDY

When you move into a new house or apartment there are number of preliminaries to be completed beforehand. If you are buying the property you will need a survey to check the building history and exclude problems such as building defects and rising damp.

---

[2] The story of David Hahn's attempts to build a nuclear reactor in his mother's shed can be found in Ken Silvestein's book *The Radioactive Boy Scout* which was published in 2004.

In some circumstances the home buyer's survey might exclude any issues with radon, but radiation would not be a common issue most buyers would consider. In 1980, a new apartment block was constructed in Building 7, Gvardeytsiv Kantemirovtsiv Street, in the city of Kramstorsk, Ukraine, in what was then the USSR (Figure 10.3).

Between 1980 and 1989 two families had lived in Apartment number 85 and over that period family members developed serious health conditions, four of whom had died from cancer. By 1989 the residents of Building 7 became suspicious that radiation might be the cause and they requested a radiation survey by a health physicist. The survey revealed that an intense source of radioactivity was located in the concrete wall between apartments 85 and 52. Part of the wall was removed and taken to the Institute for Nuclear Research at the Ukraine National Academy of Sciences where a small metal capsule containing caesium-137 was removed. It is believed that the capsule was originally part of an industrial radiation-level gauge that was lost in the Karansky quarry in the 1970s.[3] The capsule must have fallen into the gravel taken from the quarry and used in the construction of the building, finally ending up in the concrete panel that formed the wall of the apartments. Sadly the source was located in the wall of a child's bedroom, along which the bed was positioned. The dose rate measured at that point would have amounted to 18 Sv per year. On the assumption that a child might be in bed for 8 hours of every day of the year day that would have resulted in a dose of 6 Sv. The mother and two children in the first family who lived in the apartment all died of leukaemia. The eldest son of the second family died in 1987 and a younger son fell seriously ill before the source was discovered.

10.5 ENVIRONMENTAL RADIATION

Radiation is ubiquitous. It surrounds and is within all matter. Background radiation is defined by the International Atomic Energy Agency as the existing amount of ionising radiation in any particular place when no specific source of radioactivity is present. The amount of background radiation varies across the world varying with a number of factors. Humans (animals and plants) are exposed to radiation from four main natural sources: these being atmospheric and cosmic radiation, radioactivity from the earth and underground rocks, radiation emitted from the materials used to make buildings and radioactivity in food and drink. Additional radiation comes from man-made sources such as nuclear weapons fallout and from environmental releases following nuclear accidents. Globally, the annual dose of radiation received by individuals is around 2 to 3 mSv. In the UK the average annual background dose to members of the public is 2.7 mSv. The average background dose to the US population is US is 3.1 mSv. Exposures from medical radiation procedures can increase individual doses.

---

[3] Nuclear gauges are devices that use a radioactive source to measure parameters such as thickness, density, moisture or fill level. They normally incorporate a radiation source, a detector and a shutter. Gamma, beta and neutron sources can all be used in nuclear gauges and levels are measured by the attenuation of the beam by the material being loaded or manufactured. Although radiation is being emitted all the time, the device is shielded and a shutter can be closed to block the radiation beam when it is not in use.

10.6 COSMIC AND SPACE RADIATION

The sun is an obvious source of radiation but most of the radiation hitting the earth has travelled thousands of light years from sources deep in space. The high-energy particles are mainly protons, alpha particles, electrons and positrons. As these particles hit the atmosphere, they produce secondary showers of unstable, lower energy charged particles that in turn collide with other molecules in the air producing a cascade of ionising radiation falling towards the earth. Fortunately, the atmosphere provides a substantial shield that is effectively equivalent to about 4 metres thickness of concrete, so most of the radiation is absorbed. Only small amounts of low-energy particles such as protons, neutrons and electrons eventually reach the ground. Exposure from cosmic radiation at sea level is around 0.06 μSv/h, whereas at a height of 10,600 metres (35,000 feet) the cruising altitude of commercial aircraft, the dose rate is about 100 times greater at 6 μSv/h. The increased radiation dose rate from potentially harmful levels of cosmic radiation at flying altitudes has led to concern about the occupational radiation exposure to air crew. There have been a number of studies to determine if there has been any higher incidence of cancer in crew members, for example breast cancer in air stewardesses, but epidemiological studies remain inconclusive and there have been no proven links between cosmic radiation and cancer incidence.

At higher altitudes the earth's atmosphere is less dense and the magnetic field is weaker, so there is less protection against ionising particles. This represents a greater hazard for astronauts and space travellers. Space radiation originates beyond our solar system and comprises gamma and X-rays, high-energy protons and heavy ions, atoms in which electrons have been stripped away until eventually only the nucleus of the atom remains. Radiation exposure also increases during solar flare events, when particles are shot into space from the sun. NASA and all the other national space agencies have had to consider this to limit radiation exposure to crew members and have measured radiation doses to astronauts on space missions. The dose to the crew of the 9-day Apollo 14 mission to the moon was 11.4 mSv and crew members on a 6-month mission to the International Space Station orbiting at 353 km above the earth receive an average dose of 80 mSv. It is estimated that a 3-year mission to Mars would result in a dose of 1,200 mSv.

10.7 RADIOACTIVE FOOD FOR THOUGHT

Individual radiation doses are also affected by diet. Radioactivity from the earth mixes with the soil and is taken up by crops and plant food. It enters water sources and is consumed by drinking and eating fish and shellfish that take up radioactivity from sea water and the sea floor. There are two essential dietary elements that have radionuclides which add significantly to our radiation dose. These are potassium and carbon which are taken up into crops from rocks and minerals present in the soil. Radioactive potassium-40 in particular is the source in the most radioactive foods, these being nuts, bananas, potatoes, kidney beans and sunflower seeds. Whilst Brazil nuts are more radioactive than bananas, the banana has been a source of particular interest and has been suggested as an informal measure of radiation exposure. It is often difficult to convey the relative radiation exposures from different sources and many scientist and scientific communicators have used an equivalent number of days or weeks of background radiation to put the small doses into context. The amount of radiation received by eating an average-sized banana has been used as an informal measure of radiation dose which has been described as the banana equivalent dose (BED). One BED is often correlated to 0.1 microsievert (10^{-7} Sv). This can be used as an amusing way of describing small doses of radiation, but this fails when larger doses are being discussed. Given that a lethal dose of radiation for a normal person is around 5 Sv, this translates to eating around 50 million bananas. It is debatable how long this would take, but if you attempted to eat that number of bananas, death by constipation is more likely than death by radiation.

10.8 GEOLOGICAL RADIATION

The amount of background radiation from the earth varies from place to place depending on the underlying geology. In general, the terrain above granite rock formations have greater levels of background radiation. About 75% of background radiation comes from the earth either as radon gas or natural gamma radiation emitted by soil and rocks. The remaining 25% come from radionuclides incorporated in the human body through diet and from cosmic radiation. The areas in the world with the highest levels of natural background radiation include Ramsar, near the Caspian Sea in Iran, Guarapari on the southeast coast of Brazil, Karunagappalli in the Indian state of Kerala, Arkaroola in South Australia and Yangjiang in southern China. The Ramsar Talesh Mahalleh district in Iran is considered to be the most radioactive inhabited area of natural radiation on earth, due to the radioactivity in the ground, building materials and from the waters of nearby hot springs. The residents living in this district receive an average dose of 10 mSv per year. If you lie on the Guarapari beach in Brazil radiation levels could reach 175 mSv per year. Many studies have been carried out to determine if people living in these areas have higher incidence of cancer or childhood deaths than those living in areas of average background radiation. The published studies have not shown any positive correlation between cancer incidence and increasing dose rates in regions with elevated natural background radiation.

10.9 THE MOST RADIOACTIVE PLACE ON THE EARTH

Many of the sites where nuclear accidents have occurred remain highly radioactive and most have undergoing active remedial measures to contain the further spread of radioactivity. The Polygon nuclear weapons test area in Kazakhstan, described in Chapter 7, is high on the list of one the most radioactive places on earth; however, there are a number of other sites that are of equal notoriety. In the US this includes the Hanford production site in Washington and the BOMARK RW-01 Site next to the McGuire Air Force Base in New Jersey, where the premature explosion of a nuclear missile contaminated the surrounding area with radioactivity. Also in the US, the accident at Church Rock Uranium Mill in New Mexico holds the record for the largest release of radioactivity in American history. The mill operated from June 1977 to May 1982 to process uranium ore dug from mines on privately owned land 27 km north of the city of Gullap in New Mexico. On 16 July 1979 a large breach in a waste holding pond led to over 93 million US gallons of acidic, contaminated water and slurry to drain into a tributary of the Puerco River. Overall around 1.7 TBq (1.7×10^{12} Bq) of uranium, thorium, radium and plutonium flowed 130 km downstream through Gallup reaching as far as Navajo County, Arizona, blocking sewers and affecting nearby aquifers. Following the incident local water supplies were shut down and fresh water was delivered by road. The surrounding Navajo reservation was considered to be an area of low population density and requests for the contamination to be classed as a national emergency were refused. Some subsequent studies have shown that since the 1950s there have been significantly higher rates of cancer in the Navajo Indian communities.

There is one particularly notable legacy from the Russian nuclear efforts to produce materials for building weapons and power stations. When the communist leaders selected the locations for strategic facilities, they had little regard for the welfare of individuals and acted with impunity, keeping many aspects of their work a secret. This was particularly so in the sites chosen for the production of radioactivity and nuclear reprocessing. Some of the most covert activities took place at the Mayak Chemical Combine in the southern Ural Mountains, around 1,400 km east of Moscow. Construction of the Mayak facilities began in 1946 near the town of Ozyorskand, which was one of the 'plutonium cities' only known by a post office box number, in this case, Chelyabinsk-40. The Mayak Production Association site covered around 200 square kilometres and included nuclear reactor buildings, processing and fabrication facilities and a series of natural lakes, ponds and artificial reservoirs that was created by damming the Techa River. The first uranium graphite reactor became operational for the production of weapons-grade plutonium in 1948. Production continued for nearly 40 years until plutonium-239 was stopped in 1987, but two reactors remained in use for the production of plutonium-238 and the nuclear reprocessing facilities also continued to operate.

Since the 1950s there were at least six incidents at the Mayak site, some of which were criticality accidents resulting in fatalities and some resulted in the release of radioactivity and significant widespread contamination. These included the release of large amounts of radioactivity into bore holes in the ground and deliberate discharges from the waste ponds and containment reservoirs into the Techa River. Between 1949 and 1956 over 100 PBq (100×10^{15} Bq) – a huge amount – of radioactive waste was directly discharged into to the local watercourse, despite the Techa River being a main source of water for many of the villages further downstream. The radionuclides included a number of uranium and plutonium products as well as ruthenium-103 and ruthenium-106, caesium-137, strontium-90 and cobalt-60. In 1957 the cooling system in one of the radioactive waste tanks at Mayak containing about 75 tons of liquid waste developed a fault that was not adequately repaired. The temperature in the tank started to rise causing the contents to evaporate. The hot dried waste consisted of ammonium nitrate and various other nitrate chemicals. Ammonium nitrate is commonly used in agricultural fertilisers and is also an ingredient used in explosives (being the cause of the Beirut Port explosion on 4 August 2020). On 29 September 1957 the hot, dried radioactive waste exploded with an estimated force of about 70–100 tons of TNT, throwing the 150 tonne concrete lid high into the air. There were no immediate casualties as a result of the explosion, but it was estimated that around 750 PBq (750×10^{15} Bq) was released, of which

90% settled in the immediate surrounding area. The remainder of the activity 74 PBq (74×10^{15} Bq) was carried up in a radioactive plume to a height of 1 km before being dispersed by the prevailing winds and air currents. This high-level liquid radioactive waste dispersal subsequently became known as the East Urals Radioactive Trace. Approximately 10,200 people were evacuated from the contaminated areas after the accident. Over 1,000 residents in the three villages located nearest the accident site received the highest radiation doses, with average dose of 570 mSv to red bone marrow.

To address the radiological problems of the area, the Norwegian Radiation Protection Authority worked jointly with Russian radiation scientists to study the Mayak region and the surrounding populations affected by the radioactive contamination incidents. Populations exposed to radioactive contamination from the Mayak operations included the Mayak workforce, individuals involved during the clean-up work and the local residents who suffered long-term poisoning as a result of the environmental release of radioactivity. The populations living along the Techa River were particularly affected and were chronically exposed to both external irradiation and internal consumption of radioactivity. In some riverside villages the river was the only source of potable water and it was also used for bathing and washing. The water was given to animals and cattle, used for irrigating crops and breeding fish and waterfowl. In turn, the local residents received higher radiation poisoning by eating the contaminated vegetables, meat and fish and drinking the milk. External irradiation from the Techa River bottom sediments and shoreline was also an important factor contributing to radiation exposure. The radionuclides believed to have caused the most harm to health were strontium-90 and caesium-137. The average strontium-90 concentration in Techa river water sampled between July and August 2004 at Muslyumovo village, 40 km downstream from the plant, was six times higher than the Russian intervention level, meaning that living in this settlement was seen as potentially hazardous to health by the Russian Federal Medical-Biological Agency.

Scientific studies of the Techa river populations have reported an increase in total cancer deaths for residents living in the riverside communities during the period between 1950 and 1982, when compared to unexposed residents in the same region; however, there were significant differences between different ethnic groups. The absolute risk of developing leukaemia was calculated in one study to be higher than for equivalent non-exposed populations, but this was lower than that seen in the atomic bomb survivors. This was considered to be because the radiation exposure to the people living in the riverside towns and villages took place over a longer period. Significant depletion of blood cells, greater immuno-suppression and higher numbers of still births were also evident in the upper Techa residents. Analysis of 60 individuals from each of the four most exposed remaining settlements in the Techa River region also showed a significant increase in the frequency of chromosome aberrations in blood cells compared to control populations.

The Techa River drained into Lake Karachay, which when translated means 'black water'. In the 1960s the lake dried out and a regional drought left a radioactive dustbowl. In 1967 high winds carried radioactive dust to over half a million people in the surrounding villages and towns. The lake was infilled in 1978 and 1986 with around 10,000 concrete blocks to limit surface erosion. This area of the South Urals has the notoriety of being the most radioactive polluted open site in Russia and possibly in the world.

11 ☢

Biological Effects and Medical Treatment

Radiation is like the information in this book.
Whilst you are reading you are exposed to a great deal of information.
Much of it passes by you, but the bit that you remember is the dose.
Taken from the author's lecture notes.

11.1 BIOLOGICAL EFFECTS OF RADIATION

A brief history of radiation exposure and poisoning would not be complete without some more detailed explanation of the human biological effects of radiation. The first point to appreciate is the difference between radiation exposure and radiation dose. Exposure arises from the amount of something in the surroundings to which the person may become 'exposed', whereas the dose is the total amount of the thing taken in or absorbed by the body. The next point to consider is the toxic effect of the absorbed dose. The ingestion of common salt (sodium chloride) is a good example of chemical dose-related toxicity. Salt is essential for human health in small amounts, but large doses may be harmful, particularly to cardiovascular function. This comes back to the previously described scientific view that a medicine is a small amount of a poison. There is even a school of thought that this may be true for radiation. Radiation hormesis is the hypothesis that low doses of ionising radiation around or slightly above background levels are beneficial to health and stimulate cellular repair mechanisms that help to protect against disease. Although there have been scientific studies to test this hypothesis, this concept is not supported by most international scientific authorities and government bodies.

Ionising radiation has the ability to penetrate the body delivering energy directly into the tissues and potentially damaging their cells. The radiation can be from external sources or can be from internal sources that have been taken into the body or a combination of both. The first step in estimating the amount of damage a given dose of radiation will produce requires measurement of the energy absorbed from the radiation. To express the 'absorbed dose' of radiation in a substance, a special unit 'the gray' is used, which is defined as one joule of energy deposited in one kilogram of a substance. This helps in understanding the relative effects of different amounts of radiation on tissue, but it is important to note that despite the complex equations and lengthy calculations, radiation dosimetry is an inexact science when it comes to talking numbers. One gray of one type of radiation, for example gamma rays, will produce a different biological effect to one gray of another type of radiation, for example neutrons. In addition, different tissues respond differently to similar amounts of the same radiation. To meaningfully compare the potential damage caused by similar measures of radiation of different types, irradiating possibly different parts of the body, the concept of 'effective dose' is used. The overall whole body damage from radiation expressed as effective dose therefore takes account the type of radiation and the varying sensitivities of the different tissue (see Figure 11.1).

The sievert is named after the Swedish Medical Physicist, Rolf Sievert, who made a major contribution to the study of the biological effects of ionising radiation. One sievert is equal to an effective dose of one joule of absorbed energy per kilogram of tissue, which confusingly, can seem very similar to one gray! The main difference lies in the introduction of the radiation and tissue weighting factors (Figure 11.1). In fact, for a uniform whole body absorbed dose from X-rays and gamma rays the numerical values in grays and sieverts are the same, since the radiation and tissue weighting factors are both 1. However, because the absorption of alpha particles and neutrons produces greater biological damage than X- or gamma rays, the radiation weighting factors for these types of radiation are higher (Table 11.1). The biological damage also depends on the types of tissues receiving the radiation dose. Some organs such as the breast, bone marrow and the gut are more sensitive to radiation and have higher tissue weighting factors (Table 11.2).

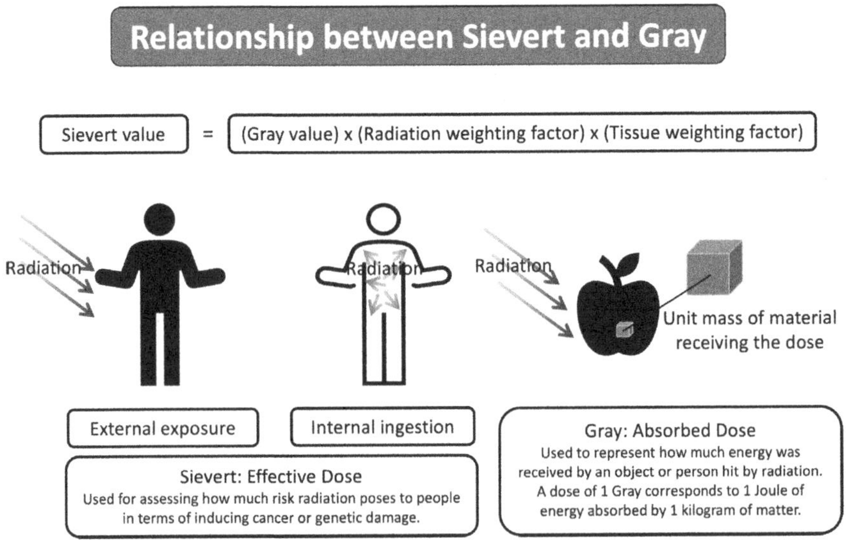

FIGURE 11.1 Exposure from external or internal sources of radiation results in an absorbed dose which is measured in grays. The effective dose in sieverts can be estimated if the types of radiation and the radiation sensitivities of the tissues affected are known.

Source: Exposure from external or internal sources of radiation results in an absorbed dose which is measured in gray. Author's material

TABLE 11.1
Radiation weighting factors

Type of radiation	Weighting factor
Photons: gamma rays and X-rays	1
Electrons: beta particles	1
Protons	2
Alpha particles, fission fragments and heavy ions	20
Neutron radiation	2.5–20

TABLE 11.2
Tissue weighting factors

Tissue/organ	Weighting factor
Breast	0.12
Red bone marrow	0.12
Colon	0.12
Lung	0.12
Stomach	0.12
Gonads (ovaries and testes)	0.08
Urinary bladder	0.04
Liver	0.04
Oesophagus	0.04
Thyroid	0.04
Salivary glands	0.01
Skin	0.01
Bone surface	0.01
Brain	0.01
Remaining organs and tissues	0.12
TOTAL	1.00

The calculation of whole body doses is quite complicated and would require some additional information such as the regions of the body affected, the size of the organs, the distribution of radiation exposure and the time course of the events. The values given in Tables 11.1 and 11.2 show which types of radiation cause the most harm and which tissues are most sensitive.

It is worth noting that both radiation and tissue weighting factors are not fixed physical constants. These are based on the available scientific data and change over time as the understanding of biological effects increases.

The detrimental health effects of radiation on the human body not only depend on the amount and type of radiation and the particular tissue receiving the dose but on the duration of the exposure. A high dose received in a short period of time will cause acute radiation effects, whereas a similar total dose delivered as lower doses accumulating over a longer period will result in chronic radiation effects. This is mainly because the cells in tissues irradiated at lower dose rates over a longer period of time have some ability to undergo repair. The absorbed energy from radiation damages tissues in two ways. These are known as direct and indirect biological processes. The sequence of events starts with the radioactive atom emitting radiation that is then absorbed by the molecules in cells and goes on to affect the whole body, as shown below.

ATOMS
Energy released and deposited affect

⇩

MOLECULES
Chemical reactions can change

⇩

CELLS
Abnormal growth or cell death affects

⇩

TISSUES
Tissue damage affects

⇩

ORGANS
Failure can affect

⇩

BODY
Resulting in death

As a radiation particle passes through tissue it does not generally lose its energy all at once in one interaction. Energy is generally lost gradually in multiple interactions along the path of the particle, leaving a track of ionised atoms and molecules behind it. Direct effects are less common since they require the radiation 'particle' to directly interact with a biological molecule, but they can cause direct damage to the DNA (deoxyribonucleic acid) in body cells. DNA is a long molecule that contains the unique genetic information for every living cell and is a prime target for biological damage. In humans, the nucleus of each cell of the body contains 23 pairs of chromosomes, one of each pair having been derived from the father and one from the mother (this number varies with different species of animals). Each individual chromosome contains two continuous strands of DNA wound around each other like a spiral staircase in the renowned double helix. An exception to this is the reproductive cells or germ cells (sperm cells in males and egg cells in females) which contain only 23 single chromosomes. Figure 11.2 shows a diagram of the DNA structure from a single chromosome.

The double helix is made up from two backbone strands linked together by sequences of the amino acids, cytosine, guanine, thymine and adenine. If incoming ionising radiation such as an X-ray, gamma ray, or an alpha particle hits the DNA it can cause a break in the DNA strand. The DNA can repair a single break on one side of the helix, since the template exists from the information on the opposite side, but if the radiation is energetic enough to rip right through the DNA causing a double strand break, for example from an alpha particle track, the information is lost and the repair will be incomplete. When this occurs, the damaged cell may die, or it

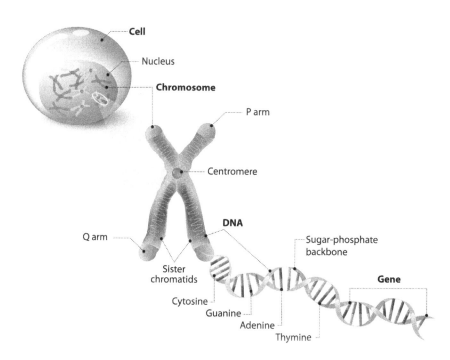

FIGURE 11.2 The chromosomes are contained in the nucleus of all living cells. The diagram shows the double helix structure of the chromosomal DNA.

Source: DNA. www.shutterstock.com/image-vector/cell-structure-dna-molecule-double-helix-368406905

may continue to reproduce in its damaged form, causing mutation or uncontrolled abnormal cell growth.

Indirect effects result from much more common products of the ionisation trail from the ray or particle and can induce chemical changes in the molecules contained in living cells. All biological systems are predominantly made up of water and the radiation-induced splitting of water molecules (radiolysis) can start the process of causing biological damage. This occurs in four stages, three of which take place quickly and one much more slowly. The sequence of events resulting in radiation damage are summarised in Table 11.3.

TABLE 11.3
The sequence of biological events following radiation adsorption in tissue

Stage	Event	Process	Duration
1	Initial interaction	Ionisation and excitation of atoms	10^{-17} to 10^{-15} seconds
2	Chemical damage	Production of free radicals	10^{-14} to 10^{-3} seconds
3	Biomolecular damage	Damage to proteins and nucleic acids	Seconds to hours
4	Biological damage	Cellular change, cell destruction or uncontrolled growth, organ failure and death	Hours to decades

The first stage is the initial physical interaction of the radiation which causes ionisation and excitation of the atoms resulting in the ejection of electrons (e-) leaving positively charged ions and molecules. Hydrogen and hydroxyl ions are normally present in living cells and do not contribute to any radiation damage. However, in the second stage of the process ions such as hydroxyl ions may be transformed into free radicals. Free radicals are electron donors (reducing species) or electron acceptors (oxidising species) which are highly chemically reactive and are capable of producing radiobiological damage. In the third stage, biomolecular damage occurs due to the free radicals reacting with organic molecules within the cell. This may be particularly harmful in the case of reactions with the DNA of the cell nucleus. The final stage of biological damage is known as the 'latent period', and this may take many years to take effect, for example in those individuals who are diagnosed with cancer many years after exposure to radiation. If the germ cells in the ovaries or testes are affected, this can result in inheritable cell changes which may be passed on to future children.

11.2 RADIOSENSITIVITY

Some cells in the body are far more radiosensitive than others, meaning that radiation will cause them far greater damage. In general, cells with faster turnover are more radiosensitive. If this were the case, it would suggest that living creatures with slower cell turnover rates would be more resistant to radiation. While a dog can only withstand up to around 3.5 Gy, a human may tolerate doses of over 5 Gy. However, there are some life forms that are far more radioresistant. Rats can withstand a dose of around 7.5 Gy and mice around 9 Gy. After the Americans dropped the atomic

FIGURE 11.3   Three of the most radioresistant creatures. Left: cockroach. Middle: Braconidae wasp. Right: water bear

Source: Left: cockroach. www.shutterstock.com/image-photo/cockroaches-on-completely-separate-white-background-1212978556 Middle: Braconidae Wasp. www.shutterstock.com/image-photo/red-braconid-wasp-family-braconidae-1388445455 Right: water bear. www.shutterstock.com/image-illustration/3d-rendered-realistic-illustration-tardigrade-1055532119

bombs on Hiroshima and Nagasaki, the only creatures that were seen to survive were cockroaches, giving rise to a belief that cockroaches would inherit the earth (Figure 11.3). In a human body some cells are continually growing and dividing, whereas in a typical cockroach the cells only divide during a period of around 48 hours a week. This gives increased resistance to large radiation doses of over 100 Gy. However, this is low compared to some insects. The ancient parasitic Braconidae wasp can withstand doses of up to 1,800 Gy but even this is no match for the radiation resistance of the *Milnesium tardigradum*, the water bear, an eight legged microanimal measuring about half a millimetre in length. This mini beast can enter a cryptobiotic state to stop its internal organic clock and halt its aging process. In 2007, experiments carried out by the European Space Agency on two species of water bear demonstrated that they could survive the near absolute zero temperatures and vacuum state in space and radiation doses of over 5000 Gy. Something that brings Ridley Scott's eponymous 'Alien' to mind.

In humans the cells turning over at the fastest rate include the haematopoietic or blood-forming cells, the white blood cells (lymphocytes), the red blood cells (erythrocytes) and the epithelial cells of the intestinal tract and the skin. Acute high doses of radiation given in a short time cause burns, blistering and loss of skin, hair loss, damage to blood vessels and sterility. From the events described in the previous chapters it is clear that the most immediate effects of high radiation exposures are reddening of the skin, gut problems such as vomiting and diarrhoea and the suppression of the bone marrow and blood cells. The other key finding from past radiation exposures is that only above certain dose levels will observable damage to specific organs occur. These are known as deterministic effects and include the immediate affects just mentioned. Deterministic effects show a well-defined threshold, below which there is no clinically observable effect and above which the severity of the effect increases with increasing dose. If data are available a dose–response curve may be drawn to describe the relationship between the dose and the severity of the effect. The formation of cataracts in the eye is an example of such a deterministic effect, although the onset is delayed. The International Commission of Radiological Protection has set a threshold dose of 0.5 Gy to the eye for cataract formation. As described in the earlier chapters the effects of radiation exposure to the skin were recognised during the early use of X-rays. It is unlikely for a skin reaction to occur at doses lower than about 2 Gy; however, above that dose the first effects take place in the capillary vessels in the dermis below the skin surface. Dilation of these capillaries and the release of histamine produce the characteristic reddening known as erythema. This may appear within minutes of irradiation and does not usually last for more than 48 hours. A second phase of erythema may occur about 1 week after exposure. This can result in the loss of superficial layers of the skin, producing a condition similar to first-degree burns. Healing may take weeks and can result in permanent scarring.

Random radiation effects, also known as stochastic effects, occur over a longer time period following significant radiation exposure. The formation of cancer (carcinogenesis), mutation (mutagenesis) and heritable effects are all random effects. For stochastic effects, there is a statistical relationship between the dose received and the probability of an effect occurring above its natural level of incidence. This means that it is the *risk* of the effect occurring rather than its *severity* which is related to the original radiation dose. It has taken a long time for scientists and clinicians to understand the random biological effects of radiation. They have been learnt by studying past exposed individuals and by carrying out long-term epidemiological studies of exposed populations.

Overall, it seems that very low levels of radiation around the average level of background are not significantly harmful to health but above very low levels, all radiation doses carry some risk. To help put the range of doses into context, Table 11.4 provides some estimated effective doses from a range of radiation sources and human endeavours. Although strictly speaking sievert values should not be used to describe doses that produce deterministic affects, this table helps to give an overall picture of the range of doses from background and various human activities.

TABLE 11.4
Examples of effective doses of radiation from background sources and critical accidents

0.05 μSv	Sleeping next to someone
0.1 μSv	Banana equivalent dose, an illustrative unit of radiation received by eating a banana
40 μSv	Air flight from New York to Los Angeles
80 μSv	Estimated average (one time) dose to people living within 16 km of the Three Mile Island accident
1 mSv	Annual dose limit for individual members of the public in the US
1.5–1.7 mSv	Annual radiation dose for air crew
68 mSv	Maximum dose to the evacuees who lived closest to the Fukushima nuclear accident
80 mSv	6-month stay on the International Space Station

160 mSv	Chronic dose to lungs over 1 year of smoking 1.5 packs of cigarettes per day, mostly due to inhalation of polonium-210 and lead-210 from tobacco
250 mSv	Radiation dose from cosmic rays for a 6-month one-way journey to Mars
670 mSv	Highest dose received by a worker responding to the Fukushima emergency
1 Sv	Maximum allowed radiation exposure for NASA astronauts over their career
4–5 Sv	Acute dose required to kill a human with a 50% risk within 30 days (LD50/30)
5 Sv	Calculated radiation dose from the gamma-ray pulse 1.2 km from ground zero of the Little Boy fission bomb
4.5–6 Sv	Fatal acute doses received to individuals during the Goiânia accident
5.1 Sv	Fatal acute dose to Harry Daghlian in the 1945 criticality accident
10–17 Sv	Fatal acute dose to Hisashi Ouchi who was kept alive for 83 days after the Tokaimura nuclear accident
21 Sv	Fatal acute dose to Louis Slotin in the 1946 criticality accident
36 Sv	Fatal acute dose to Cecil Kelly in 1958 causing death within 35 hours
54 Sv	Fatal acute dose to Boris Korchilov in 1961 during unshielded work on the failed reactor cooling system in the Soviet submarine K-19
64 Sv	Nonfatal dose to Albert Stevens after he was injected with 130 kBq plutonium-238 and 1.7 kBq plutonium-239 in 1945. The activity decayed slowly resulting in a slow cumulative dose delivered until his death 20 years later

11.3 TREATMENT OF RADIATION CASUALTIES

Radiological medical emergencies are relatively rare compared with other casualties, therefore many doctors will not have a great deal of experience in how to deal with the victims of radiation exposure. Past casualties have mainly arisen as a result of industrial and military accidents, or from exposure due to misplaced or stolen sources of radioactivity. However, it is considered more likely that future casualties may occur because of terrorist activities such as sabotage, deliberate poisoning or a radiological dispersion device (dirty bomb). Radiation casualties may have been harmed by external radiation exposure, internal contamination or both. On admission to the casualty department the first principle of care is that the standard medical and surgical needs of the patient are dealt with before decontamination and exposure management. To provide the right medical care it is helpful for the medical team to have an assessment of the radiation dose that the victim has received. This can be estimated from information available from the site where the accident occurred or by clinical means. An appropriately trained physician can estimate the severity of the radiation exposure by clinical examination to look for signs and symptoms. Many of the signs are transient, such as skin erythema, disorientation, nausea and vomiting and diarrhoea. In the case of acute whole body exposure the main symptoms are shown in Table 11.5.

Measurement of the blood cell count is important and the level of blood lymphocytes, a type of white cell of the immune system, is one of the best indicators of the severity of radiation injury. A decrease in lymphocyte counts occurs promptly within 24 hours after radiation exposure. Radiation dosimetry may also be estimated based

TABLE 11.5

Medical findings following acute radiation exposure. Taken from Koeng et al. *Annals of Emergency Medicine.* 2005;45:643–652

Symptoms and medical response	Level of acute radiation syndrome and whole body dose (Gy)				
	Mild 1–2 Gy	Moderate 2–4 Gy	Severe 4–6 Gy	Very severe 6–8Gy	Lethal >8Gy
Onset of vomiting	2 h after exposure	1–2 h after exposure	Within 1 h after exposure	Within 30 min after exposure	Within 10 min after exposure
Incidence %	10–50	70–90	80–100	100	100
Onset of diarrhoea	None	None	Mild 3–8 h	Heavy 1–3 h	Heavy within 1 h
Incidence %			<10	>10%	100
Onset of headache	Slight	Mild	Moderate 4–12 h	Severe 3–4 h	Severe 1–2 h
Incidence %			50	80	80–90
Body temperature	Normal	Increased 1–3 h	Fever 1–2 h	High fever <1 h	High fever <1 h
Incidence %		10–80	80–100	100	100
Medical management	Outpatient observation	General hospital admission, for transfer if necessary	Treatment in specialised hospital	Treatment in specialised hospital	Palliative treatment of symptoms only
Prognosis	High probability of survival	Probable survival	50% probability of death within 6 weeks	Probable death within 2–3 weeks	Probable death within 1–2 weeks

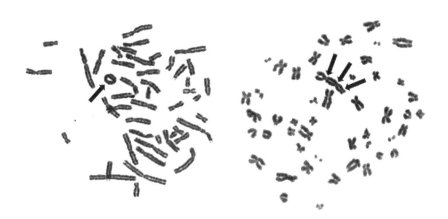

FIGURE 11.4 Examples of chromosome aberrations. Left: arrow indicates acentric ring; right: arrows show tricentric chromosome aberrations.

Source: Examples of chromosome aberrations. Author's material

on chromosome analysis. A blood sample can be taken and the cells put in culture to stimulate the lymphocytes to grow and divide. Once they reach the point of cell division known as the metaphase stage, the chromosomes can be seen under a microscope (Figure 11.4).

The damage to the chromosomes can then be identified and the type and number of defects, known as aberrations, can be related to the radiation dose. Retrospective dosimetry may also be estimated by measuring any activated radioactivity in materials and objects carried or in close proximity to the exposed individual (e.g. from neutron activation). Personal metal objects on the victim such as rings, keys, coins or a mobile phone can be used. The analysis of any neutron-activated sodium-24 produced from stable sodium-23 in the body can also be carried out. Sodium-24 has a physical half-life of 15 hours and produces beta and gamma rays with energies of 1.37 and 2.75 MeV. The activity of sodium-24 in the victim's body may be assessed directly by whole body counting or by measuring the activity in vomit or a blood sample using a gamma-ray spectrometry.

11.4 TREATING RADIATION CASUALTIES

Once a radiation casualty has been confirmed the options for treatment are limited. In situations where the victim is contaminated with radioactivity it is necessary to remove clothing and wash any external contamination from the skin. If the victim can be treated immediately after the radioactivity has been swallowed, the radiation dose can be limited by removing radioactivity from the stomach, thereby preventing absorption by the small bowel. In the past this was undertaken by gastric lavage and pumping out the stomach contents. It is also possible to induce vomiting by giving an emetic agent that causes gastric irritation and sickness. Emetics such as salt or mustard solution, or syrup of ipecac have been used in the past, but since the administration of emetics can be dangerous, oral dosing of activated charcoal is often used to bind the poison, causing it to pass through the bowel and preventing absorption into the bloodstream.

If the patient has been exposed to radioactive iodine such as iodine-131, a thyroid-blocking agent should be given to reduce the internal uptake of radioiodine in the thyroid gland. Potassium iodide is the drug of choice and should be administered within a few hours to be effective. If a metal radioactive poison has entered the circulation, it may be possible to give an agent to bind the metal and increase the speed of elimination from the body. The more quickly a radioactive poison is removed, the lower the radiation dose and the less serious the health effects will be. Prussian blue (ferric hexacyanoferrate) will increase the elimination of caesium and thallium radionuclides from the bowels. This has been used to treat casualties in the past and is considered to be a good standby in possible terrorist incidents using stolen or abandoned radiation therapy sources or industrial sources containing caesium-137. Chelating agents are also used as scavengers of radiometals and are most effective when they are given shortly after radioactive materials or poisons have entered the body. During the Second World War, dimercaprol, an organic dithiol compound, was developed as an experimental antidote against the arsenic-based poison gas lewisite. After the Second World War, lead poisoning became a major problem in naval personnel as a result of their jobs of repainting the hulls of ships. The medical use of a salt of ethylene diamine tetraacetate (EDTA) known as sodium calcium edetate was introduced as a therapeutic chelating agent to treat lead poisoning. In the 1960s a further compound meso-2,3-dimercaptosuccinic acid (DMSA), a related dithiol with far fewer side effects, was introduced. A further chelating agent diethylenetriaminepentaacetate (DTPA) was introduced and found to be a more powerful metal sequestering agent. DTPA comes in two forms: calcium (Ca-DTPA) and zinc (Zn-DTPA). In 2004 the US Food and Drug Administration determined zinc-DTPA and calcium-DTPA to be safe and effective for the treatment of internal poisoning with plutonium, americium or curium. When given within the first day after internal contamination has occurred, calcium-DTPA is about 10 times more effective than zinc-DTPA, but after 24 hours have passed, calcium-DTPA and zinc-DTPA are equally effective in chelating these radioactive materials.[1]

11.5 BEER AND DRAGON'S BLOOD

Given the limited options it is understandable if the victim might just accept that all is lost and resort to a stiff drink. In the 1957 post-apocalyptic novel *On the Beach* by Nevil Shute, the inhabitants of Melbourne, Australia, wait as a deadly cloud of radiation spreads south following the atomic obliteration of the Northern Hemisphere. The Australian government issues free suicide pills and injections, so that people can avoid the prolonged suffering from radiation poisoning. Towards the end of the

---

[1] These metal chelators are used in medical diagnostic radiopharmaceuticals and image contrast agents. For example, technetium-99m-DTPA is used in nuclear medicine for the dynamic imaging of kidney function and gadolinium-DTPA is used for delineating tumours by magnetic resonance imaging.

book a scientist considers if alcohol might offer 'immunity from radioactive disease'. A similar reference to the protective properties of alcohol is made in the submarine movie *K-19 the Widowmaker* when the captain tells the crew to drink vodka to help protect them against the radiation. Whilst drinking may induce some degree of calming it is unlikely that alcohol would have any real affect. However, since one of the secondary effects of radiation poisoning is the production of free radicals, it interesting to note that alcohol is in fact a scavenger of free radicals. In 2002, two scientists from Japan published a scientific study on the effects of beer drinking on radiation-induced chromosome aberrations in blood lymphocytes. Human blood was collected before and after drinking 700 millilitres of beer and the samples were irradiated with 4 Gy X-rays or 4 Gy of carbon ions. Chromosome analysis of the cultured lymphocyte samples demonstrated that the number of aberrations was significantly reduced in the blood taken after beer had been consumed. Perhaps there may be some merit in resorting to the bottle, although in reality alcohol poisoning may be a more likely outcome.

There is also one other intriguing scientific publication claiming that dragon's blood has a radioprotective effect after exposure to gamma rays and heavy ions. Whilst this might seem to be from the realms of pure fantasy, it is a serious scientific study that was published in the *Journal of Radiation Research* in 2012. After further scrutiny of this work, it would appear that dragon's blood is a resin obtained from *Dracaena Cochinchinesis*, an evergreen tree-like plant found in Asian tropical forests and otherwise known as the dragon tree. The bright red resin taken from the plant is used as a traditional Chinese medicine to promote wound healing and to stop bleeding. Chinese researchers showed that rats injected with dragon's blood before being exposed to whole body irradiation with gamma rays and heavy ions showed significantly lower levels of biochemical markers associated with radiation-induced damage and inflammation than similarly irradiated control rats.

The previous chapters have given historical accounts of the treatment of acute radiation victims, including the intensive treatment of Hisashi Ouchi, which led to increased suffering rather than providing compassionate care. The medical management of acute radiation exposure includes treating injuries and burns, reducing infections and maintaining hydration. Some patients require bone marrow rescue to maintain the production of blood cells. If patients suffer·from persistent marrow failure, there are drugs which can be given to simulate marrow function and allogeneic stem cell transplantation can also be attempted to re-establish the normal level of circulating blood cells.

In those that survive, the recovery process may last from weeks up to years after the dose was delivered and they may eventually die as a result of the delayed effects of radiation-induced cancer. The past experiences of dealing with radiological incidents has also shown that there are also wider effects on many people. The psychological aspects of treatment should be considered, not only for the severely affected victims but for the wider population. In any significant radiological event there will always be large numbers of 'worried well' who need psychological support and reassurance.

12 ☢

Spies, Subterfuge, Missions and Murder

Killed by a little, tiny nuclear bomb. . .
Comment made to the media by Walter Litvinenko after the death of his son
Alexander

12.1 SPY DUST

In the 1969 James Bond movie *On Her Majesty's Secret Service*, there is a scene where James Bond discusses the obsolete nature of the old special equipment with Q, the quartermaster and head of research and development. Q goes on to reveal new miniaturised hi-tech devices designed by computer. One such new approach was radioactive lint, which could be placed in the opponent's pocket for anti-personnel location. This may seem rather far-fetched, especially as the enemy agent might decide to change his jacket or simply throw the scrap of material in the bin with a toffee wrapper. The fictional character of Agent 007 of the British Secret Service was originally conceived as a Cold War spy operative by the British novelist Ian Fleming who himself served in the British Naval Intelligence Division. Many of 007s exploits seem far-fetched but in reality, the truth is often stranger than fiction. At the end of the Second World War the German city of Berlin was in ruins and only about 2.8 million of the original population of 4.3 million people remained. The winning Allied countries of America, Great Britain, France and the Soviet Union divided the city into 4 sectors. After 1946 escalating conflict between the East and West led to Berlin becoming the centre of Cold War conflict. In 1948 the Soviet Union blocked all the roads, railways and canals around the western sectors of the city in an attempt to force the American, British and French forces out, preventing food supplies from entering the city. The Western Allies responded with the Berlin Airlift, dropping around 2.3 million tons of goods for over a year. The Soviet blockade was a failure, but the Soviets retained influence over East Berlin and on 7 October 1949 when the German Democratic Republic (GDR) was founded in the Soviet sector, East Berlin became the capital.

On 8 February 1950, Wilhelm Zaisser, the first Minister of the Ministry for State Security of the GDR, established the State Security Service (*Staatssicherheitsdienst*, SSD) which was formed in the same style as the KGB and became known as the 'Stasi'. The Stasi's emblem was the 'shield and sword of the party' and its sole purpose was to keep the Communist Party in power by any means possible. If someone looked like he might challenge the party's legitimacy or control, the Stasi systematically destroyed his or her life. They did this by surveillance and control, including intimidation, harassment, threats, blackmail, denying education and employment, and by imprisonment and torture. The extent of surveillance and distrust caused the East German population to live in terror. At the peak of the 40-year Communist control of East Germany the Stasi had 91,000 employees, meaning about one in every 30 residents was a Stasi agent. More than one in three of the population (5.6 million people) were under suspicion. Their homes were searched, their phones were bugged, their letters opened and copied, and their movements secretly filmed or photographed. Every document went into a personal Stasi file. The invasive and pervading nature of the surveillance operations were portrayed with great authenticity in the 2006 German film *The Lives of Others* (Das Leben der Anderen). Since German reunification in 1990 the true extent of the records held has been realised. Hundreds of millions of files, 39 million index cards, 1.75 million photographs, 2,800 reels of film and 28,400 audio recordings have been recovered from the Stasi Records Agency. As the Berlin wall fell the order came to destroy the records. The officials made every attempt to shred the files but there was so much paper that the shredders failed. They resorted to burning and pouring chemicals over them and as last resort tearing them by hand, but the task proved to be too great. The people stormed the Stasi buildings and recovered 16,000 sacks of torn-up paper. The process of reconstructing the partially destroyed files has been likened to completing a huge jigsaw, which it has been estimated will take 400 years to complete. The German Federal Archives now hold these records and in the interests of openness are allowing citizens to access those files that can be read. The surviving documents have revealed a picture of how the Stasi used technology for surveillance and monitoring. A fascinating account of the Stasi's spy-tech world was described by Kristie Macrakis, a Professor from the School of History, Technology and Society at Georgia Institute of Technology in Atlanta, in a book published in 2008 titled *Seduced by Secrets*.

The Stasi security services used as much modern technology as they could get to monitor the conversations of citizens and the movement of individuals. A key motivation for the extent of the surveillance was to detect subversive activities such as spying and espionage and to identify potential defectors and the movement of top-secret documents. The Stasi Technical Operations Sector was one of the largest Divisions of the Security Ministry and occupied buildings in a restricted area next to the infamous Hohenschönhausen, Stasi prison. These departments worked on invisible inks, miniature spy cameras (mikrats), tape recorders, radio bugging equipment, code tables and coding devices and the deployment of dyes, chemicals and radioactivity. From this building the Stasi secret agents would be trained and issued with equipment for their missions, in the same way that James Bond stories show Q issuing secret gadgets to double 'O' agents. Spy cameras were commonly used to steal secrets and to miniaturise information for delivery elsewhere. These were small devices that could be hidden in objects such as a cigarette case, or positioned behind the button of a jacket, or in the cleavage between the cups of a bra. Once the photographs had been taken, the film would be transported in a self-destructive container, such as a hair spray or baby powder canister. The container also contained a battery and flash bulb. If an unintended person attempted to open the container in an incorrect manner, the bulb would flash destroying the information on the film. In 1960, Alfred Frenzel, a West German member of Parliament who worked for the Czech services, was caught using such containers.

The Stasi went to a great deal of effort devising ways of marking items for tracking purposes. Marking objects to identify if someone had stolen them has been common practice dating back hundreds of years. Marking money with something identifiable, such as scratching coins or making pencil marks on notes or documents, is perhaps considered to be one of the first applications of a scientific method to capture a thief. The first individual to have made a formal study of the use of scientific aids in criminal investigations was Professor F. G. Tryhorn of the University of Hull in the UK. He is credited to be the founder of forensic science and carried out work on the analysis of dust particles to aid criminal detection. He is also credited to be the first person to consider the use of radioactivity in criminal investigations. A paper titled 'Identification of Objects by Radioactive Labelling' published in 1940 by Tryhorn and Widdowson, in the *Police Journal: Theory, Practice and Principles*, discussed the relative merits of labelling items with dyes, chemicals and radioactivity. The authors did not, however, consider the health implications of using radioactivity. Other organisations subsequently experimented with the labelling of items with dyes, chemicals, fluorescent materials and radioactivity. There is evidence that the FBI experimented with radioactive tracers to apprehend criminals in the 1940s. The Stasi programme of work to develop secret methods for tracking items and people was given the project name 'Cloud'. Dr Franz Leuteritz a nuclear physicist, who had been studying at the Academy of Sciences Institute for Applied Radioactivity in Leipzig, was appointed in 1967 and soon became the head of the Cloud project. Tracking methods included dyes and fluorescent chemicals that would glow when examined under ultraviolet light. Three main methods for tagging items with radioactive markers were developed: a self-adhesive radioactive plastic foil, radioactive needles to be pinned on materials such as fabrics and radioactive liquids to be applied using a spray. A wide range of radionuclides were used for the Cloud missions. These were clearly supplied by the Soviet nuclear manufacturing facilities, most likely the Rossendorf Nuclear Reactor Institute, which begs the question, did the Soviets deploy these radioactive tracer methods at any time in the past? A list of the various radionuclides and applications used by the Stasis has been obtained from the surviving shipment documents, dating from 1972 to 1989. These have allowed the estimation of the type and quantity of the deliveries made which are summarised in Table 12.1.

The radioactivity was contained in specially designed vials that could be sealed and opened using a magnetic tool (Figure 12.1)

For the purpose of pursuit and detection the Stasi agents would use liquid radioactivity such as scandium-46 to mark papers, documents, letters, manuscripts, passports, bank notes, ribbons and ballpoint pens. Clothing, wallets, bags, suitcases, briefcases and cigarette cases were also common items labelled with a needle, pin or foil containing radioactive silver-110m. To mark cars the agent would attach a radioactive magnet to the underside of the body. For marking cars and vehicles in motion, the Cloud technical team developed an airgun that could fire an aluminium projectile containing silver-110m wire. This was aimed at the tyre of a

TABLE 12.1
Spy dust methods used in the Stasi 'Cloud' program in the 1980s. Data taken from 'Seduced by Secrets, Cambridge University Press' with permission from Kristie Macrakis, Georgia Institute of Technology, Atlanta

'Cloud' method	Radionuclide	Half-life
Plastic		
47100–010	Mn-54 (manganese)	312.2 days
47100–020	Sc-46 (scandium)	83.81 days
47100–030	Co-58 (cobalt)	70.8 days
47100–040	Ba-140 (barium)	12.7 days
Needles		
47100–310	Fe-59 (iron)	44.5 days
47100–320	Co-58 (cobalt)	70.8 days
47100–330	Ag-110m (silver)	249.8 days
Liquid Spray		
47100–650	Br-82 (bromine)	1.4 days
47100–670	Co-58 (cobalt)	70.8 days
47100–680	Cs-137 (caesium)	30.2 years
Paper		
47101–310	Mn-54 (manganese)	312.2 days
47101–320	Sc-46 (scandium)	83.81 days
47101–330	Ba-140 (barium)	12.7 days
47100–340	I-131 (iodine)	8 days
47100–350	Br-82 (bromine)	1.4 days
47100–360	Na-24 (sodium)	4.96 hours
47100–370	Co-58 (cobalt)	70.8 days
47100–380	C-14 (carbon)	5715 years
47100–390	Cs-137 (caesium)	30.2 years
47100–400	H-3 (hydrogen)	12.3 years
Pen Ink		
47101–610	S-35 (sulphur)	87.2 days
47101–620	P-32 (phosphorus)	14.3 days
47101–630	Pr-143 (praseodymium)	13.57 days
Car Magnets		
47102–10	Sc-46 (scandium)	83.8 days
47102–20	Co-58 (cobalt)	70.8 days
47102–30	Nb-95 (niobium)	4.97 days
47102–40	Ba-140 (barium)	12.7 days
Air Gun		
47101–010	Ag-110m (silver)	49.8 days

FIGURE 12.1   Glass vial used for the transportation and storage of the tracer radionu-clides. The vial was sealed with a magnetic cap which could be opened and closed with a hand-held magnetic tool.

Source: Glass vial used for the transportation and storage of the tracer radionuclides. The vial was sealed with a magnetic cap which could be opened and closed with a hand-held magnetic tool. With thanks to the BStU archive, Berlin and Kristie Macrakis, Georgia Institute of Technology, Atlanta

car where it would penetrate the rubber and remain in place for subsequent detection. When the vehicle was stopped at a security point, the security guards would check for the presence of any radioactivity under the bodywork or in the tyres. For detection of the gamma radiation emitted from the tracer sources, a total of five radiation monitors were developed. These used scintillation detectors because they had greater sensitivity than Geiger counters for detecting the small amounts of radioactivity in sprays, needles and foils. The instruments were code-named 'Cloud 001' through to 'Cloud 005'. The 'Cloud 005' was the most advanced, since this could be hidden under the security agent's jacket (Figures 12.2 and 12.3). A harness that fitted across the shoulders supported the scintillation detector and electrical battery pack. A wire leading from the instrument was connected to a vibrator that was placed close to the skin. If radioactivity was detected, the pad vibrated giving a silent alert, similar to the way a radiopager or modern mobile phone works today. Using such a device, the agent remained inconspicuous and could then choose his moment to apprehend his target.

Some interesting cases describing how the Stasi used radioactive trace equipment have survived and have been written up by Hermann Vogel, who was head physician

FIGURE 12.2   The Cloud 005 scintillation detection instrument on the left was powered by a battery pack. The small black vibrator pad worn near the skin of the agent can be seen on the bottom right.

Source: The Cloud 005 scintillation detection instrument. With thanks to the BStU archive, Berlin and Kristie Macrakis, Georgia Institute of Technology, Atlanta

and professor of radiology at the Albers-Schönberg-Institut, St. Georg Hospital in Hamburg, Germany.[1]

From Vogel's scrutiny of the documents, in 1975 alone, Stasi agents carried out over 100 tracked pursuits with radioactive substances, 40 using marked papers, 42 using foils, 25 using sprays and 6 with radioactive needles. Records also showed that between 1982 and 1989 there were 60 missions that included the use of radioactive substances, about half of which involved marking of paper sheets with scandium-46. The specific cases listed included a design engineer who was under suspicion of passing information to a West German contractor. By labelling four documents with radioactivity, the Stasi traced the location of the radiation, proving that the engineer stole the documents and took them home, before passing them on to someone from West Berlin. Each document was tagged with approximately 17 MBq (450 microcuries) giving a total activity of 67 MBq (1.9 millicuries). It is highly likely that scandium-46 was used as the tracer which decays giving off high-energy gamma rays of around 1 MeV. Using the Cloud scintillation detectors, the Stasi agents claimed that they could detect gamma rays of

---

[1] During his career Professor Hermann Vogel promoted the use of forensic radiology and created a digital archive of over 1,500 X-rays of the victims of violence, mutilation, torture and terrorism obtained during visits he made to countries such as Vietnam, Mexico, Zimbabwe, South Africa, Chile, Croatia and Turkey.

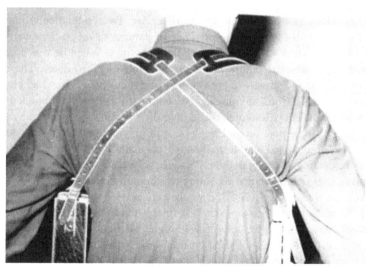

FIGURE 12.3   A leather harness for wearing the radiation detector and its battery pack.
A vibrator positioned under the armpit was used to alert the operator when radioactivity was
detected.

Source: A leather harness for wearing the radiation detector and its battery pack. With thanks to the
BStU archive, Berlin and Kristie Macrakis, Georgia Institute of Technology, Atlanta

this energy through the door or walls of the engineer's apartment. The engineer and his
wife were subsequently arrested for stealing and passing on the documents.

In another case the Stasi had obtained information to suspect that a pair of identical female twins intended to exchange passports so that one twin could escape to the
West, where the other was already living. The Stasi planned to mark the sister living in
the East with radioactivity so that they could detect her as she crossed the border. Two

agents were trained and issued with the radioactive spray, but the plan was abandoned, because the security services had found a way to identify each twin by other means.

Rudolf Bahro was a writer and dissident who lived in East Germany but managed to get a substantial piece of work published in the West. Since his writing was regarded as subversive by the Communists, he had to hide his work away from state officials. On one occasion in 1977 he hid a 321-page manuscript outside his house with the intention of sending it in separate parcels to sympathisers in the GDR, then on to other neighbouring eastern bloc countries and eventually further on to West Germany. The Stasi found his hiding place and sprayed radioactivity on the sheets of paper, so that they could track their journey and identify the individuals who were helping him. By monitoring the gamma signals, between 23 and 24 August, the Stasi traced 17 mailings which provided evidence as to who was handling the written material. The Cloud method had identified Bahro's sympathisers, couriers and friends who were all imprisoned. In October 1979, the 30th anniversary of the founding of the GDR, Bahro was granted amnesty and was deported with his family members.

A further case showed how radioactivity was used to identify the theft of money. On 4 May 1988, Stasi operatives labelled 25 East German Deutsche Marks (DM) with a total of 2.2 MBq (60 microcuries) of silver-110m. The notes were placed in a sealed envelope and posted into the mail service. By systematically monitoring the mail staff at a distance for the presence of gamma radiation they were able to identify one individual who was showing high readings. When he was approached and searched, he was carrying 8 of the 20 radioactive-tagged notes. The other 12 DM were never found. It was interesting that the security services had considered the radiation doses to the individuals who continually carried the radioactive notes around with them. They estimated that a dose of 2 Sv could be received by continually carrying a single note around in a pocket. Carrying multiple notes could put a person at risk of localised long-term damage, particularly if the money was close to the testes or ovaries in young people. The dangers were also apparent in the case of radioactive notes being handled by young children or pregnant women. To reassure operatives, the Stasi issued a list of what they considered safe radiation doses to the pursuing agent and the target suspect. It is interesting to note that the doses considered acceptable for the target suspect were higher than those considered acceptable for the agents; however, it should be borne in mind that the Stasi agents would be receiving an occupational dose which would be repeated during ongoing surveillance work (Table 12.2).

12.2 X-RAY SURVEILLANCE

The population of East Germany were not allowed to leave the county and escaping was considered a crime punishable by imprisonment or death. The GDR used both gamma rays and X-rays in the pursuit of dissidents and to detect individuals who were trying to escape. To search for hidden items and spaces the Stasi used radioactive sources and X-rays to carry out radiological inspection of apartments and houses. X- and gamma radiography was used to detect equipment hidden in walls. The use of gamma rays in particular would result in considerable exposure to individuals in apartments or adjacent premises. At border crossings the Stasi technical

TABLE 12.2
Schedule of acceptable radiation doses for security agents and target suspects set by the GDR Security Ministry for operations during radioactive tracing operations

Distance from the body (cm)	Permissible dose over time	Location of radioactivity carried	Person
3	<1 mSv/week	Trouser pocket	Stasi agent
3	<3 mSv/week	Coat pocket	Stasi agent
30	<1 mSv/week	Briefcase	Stasi agent
3	<1 mSv/week	Trouser pocket	Target suspect
3	<80 mSv/action	Coat pocket	Target suspect
30	<80 mSv/action	Briefcase	Target suspect

team investigated the use of caesium-137 sources suspended in gantries over roads, using a collimated beam of gamma rays to carry out the surveillance of passing vehicles. As a car passed under the source, a collimated fan-shaped beam of gamma rays directed towards a detector line was used to detect people hiding under seats and in hidden compartments. Unknown to western countries the use of low-energy X-rays for security scanning of luggage was introduced early on at the East German Border crossing points to locate secret compartments and other items in bags and cases. At one point the Stasi scientists carried out research into the use of X-ray-safe containers which would not reveal their contents or be mistaken for other innocent items. This work was not successful and was not widely used. Security search by X-rays is now used routinely to assist in the detection of weapons, drugs, contraband and illegal immigrants in lorries. X-ray fluoroscopy, backscatter imaging and computed tomography are used to search packages and luggage in transit, and low-energy X-ray techniques are used to examine suspected 'body packers' to discover drugs transported inside the body, on the body surface and in clothing.

There is one other uncertain aspect concerning the use of external irradiation by X- and gamma radiation by the Stasi. A number of notable dissenters who were imprisoned by the GDR include Rudolf Bahro Juürgen Fuchs and Gerold Pannach, who subsequently died of malignant disease. After the fall of the Berlin wall and the Stasi's use of radioactivity was confirmed, speculation also arose as to whether these malignant diseases were the results of intentional surreptitious exposure to radiation performed during imprisonment. Although a number of X-ray units were found in the Stasi prison, there was no conclusive evidence that this equipment was used to deliver lethal doses of radiation to prisoners.

12.3 THE LONDON POLONIUM POISONING

At 2 am on 1 November 2006, the London ambulance service responded to a 999 emergency call concerning a 43-year-old man called Edwin Redwald Carter who was taken ill at his home at 140 Osier Crescent, Muswell Hill, North London. The man had been up all night with severe stomach pain and had been vomiting violently. Two paramedics attended and after examining him and taking his blood pressure and pulse, they concluded that he had some kind of bacterial infection and told him to him to take water and pain killers. Throughout the following day Carter's pain became worse and he began to suffer from bloody diarrhoea. Early in the afternoon a doctor from the family medical practice was called. The doctor felt that the symptoms were similar to typhoid fever, but he thought that it was something more serious and called for an ambulance to take him to hospital. On 3 November, Mr Carter was taken to Barnet Hospital, where he was admitted with suspected gastroenteritis and mild dehydration. He was put in a ward side room and was given intravenous fluids and treated with 550 milligrams ciprofloxacin, a broad spectrum antibiotic, every 12 hours. Although he was feeling a little more comfortable the doctors were concerned that his blood white cell and platelet counts were decreasing, something that would not occur in a straight forward case of gastroenteritis. The other odd aspect of this patient was that Mrs Carter had told the doctors that her husband had worked in Russia for the KGB and knew some dangerous people who may have poisoned him. As might be expected, on first hearing this the medical team thought the story was rather far-fetched, but they recorded it in the medical notes on 7 November. Carter and his wife asked the doctors whether poisoning by infection with *Clostridium difficile* (a bacteria that can infect the bowel) might have occurred, since they had a friend who had been killed in this way. He was given oral metronidazole (400 milligrams three times daily) but his gastrointestinal symptoms persisted. By 9 November his white cell and platelet counts were dropping further and he had spiked a fever. Things took a turn for the worse on 11 November when Carter's hair began to fall out and on the next day he started complaining that his throat was hurting. On 13 November Dr Andres Virchis, a haematologist at the Barnet Hospital, was called to assess the patient. He thought the signs and symptoms were more like those of cancer patients who were being treated with cytotoxic drugs for tumour chemotherapy. Carter continued to insist that he had been poisoned and had been contacting the news media services including the BBC Russian Service. After reviewing the case Dr Virchis considered that poisoning may be the only explanation. He decided to check for any radiation contamination and asked the Radiology Department to check for any evidence of radiation. Checks with a scintillation detector would have only picked up external gamma rays or high-energy beta radiation and these proved to be negative. On 14 November, he contacted the Poisons Unit at Guy's Hospital in London which undertook a biopsy and tested for heavy metal poisoning. The results came back with a provisional diagnosis of thallium poisoning. A bone marrow sample taken on 15 November showed that his blood-forming cells were damaged and lacked normal cell structure. At 5:30 pm on 17 November the doctors wrote in Carter's medical notes:

The patient has developed headache and rigors (spiking temperature and chills). Not able to check for possible radioactivity. Await assays for other heavy metal poisoning. Possible need for bone marrow transplant at UCH (University College Hospital) when a bed is available. Plan:

1. Transfer to UCH.

2. Continue Prussian Blue 4 g t.d.s. (three times a day) for 7 days until repeat urinary thallium level available

3. Continue present antibiotics

Later that evening Carter was transferred to the Critical Care Unit on the 16th floor of University College Hospital, London, where he was placed under guard by armed police. Shortly after midnight two detectives from Scotland yard arrived to question him. They soon learnt that Edwin Carter had a previous career in the Russian security service and had escaped from Russia. He had been granted political asylum and citizenship in the UK and was working for the British Secret Intelligence Service, MI6. His real name was Alexander Litvinenko (Figure 12.4).

Alexander Litvinenko was born in the Russian city of Voronezh in 1962. After finishing school in 1980 he took a position as an Interior Troop Private in the Russian Ministry of Internal Affairs, a Russian interior police organisation. After a series of moves and promotions, in 1997 he was appointed as a senior operational officer in the Russian Federal Security Services (FSB) working in the Directorate for Suppression of Criminal Groups. In 1994 he became acquainted with Boris Berezovsky a Russian oligarch who had survived an attempted assignation. After a promotion to the FSB Directorate of Analysis and Suppression of Criminal Groups, he uncovered a network of corruption with connections between the Russian mafia and senior members of the Russian Law enforcement agencies. Litvinenko reported his findings to

FIGURE 12.4   Alexander Litvinenko.

Source: Litvinenko. www.shutterstock.com/editorial/image-editorial/britain-berezovsky-london-united-kingdom-7149596a

the Director of the Federal Security Services who at that time was Vladimir Putin. Putin was not interested. Litvinenko and three other officers decided to go public and held a press conference at the Russian News Agency. As a result Litvinenko was dismissed from his position and arrested. After his release he was told not to leave Moscow but he travelled with his family to Turkey and applied for asylum in the US. His application was denied. He then took a flight from Istanbul to Moscow via London and sought asylum between connecting flights at Heathrow Airport. His request was granted by the British authorities and he worked as a journalist and for a while lived in Boston, Lincolnshire, where he wrote two books: *Blowing up Russia: terror from within* and *Lubyanka Criminal Group*. In his books he accused the Russian secret service of staging the bombing of a Russian apartment building and of other terrorism acts. He also accused Vladimir Putin of ordering the murder of the Russian journalist Ann Politkovskaya in October 2006.

Litvinenko had settled into his life living and networking with various contacts and other Russian exiles in London and Europe. He was a part time salaried informant for MI6 and had been issued with a UK passport and an encrypted mobile phone. On 1 November 2006, his past caught up with him when he travelled into the Mayfair district of London. He had taken a bus from Muswell Hill and the tube to Oxford Circus to meet an Italian associate Mario Scaramella for a late lunch at the Itsu Sushi Restaurant, 167 Piccadilly, at 3 pm. During lunch Litvinenko received documents stating that the journalist, Anna Politkovskaya, had been the victim of a contract killing. On 7 October 2006 she was found dead in the lift of the block of flats where she lived in central Moscow. She had been shot twice in the chest, once in the shoulder and once in the head at point-blank range. At around 4 pm, Litvinenko walked to Grosvenor Square and met two former Russian agents Dmitry Kovtun and Andrei Lugovoy at The Millennium Hotel (Figure 12.5).

Litvinenko met Lugovoy in the hotel foyer and they went into the Pine Bar, which is immediately to the left of the main entrance. Lugovoy had already ordered drinks and they sat at a corner table with Kovtun. This was the second time that Litvinenko had met Kovtun, but he later commented that he seemed tense and there was something strange about him. The meeting only lasted for 10–15 minutes. Litvinenko refused the offer of a drink but took half a cup of green tea from the pot that was already on the table. After leaving the Pine Bar, Litvinenko went to the office of the Russian Oligarch Boris Berezovsky who was by now also living in the UK.[2] He used a fax machine and another associate offered him a lift him home to Muswell Hill in his car. It was several hours after these meetings that Litvinenko started to complain of feeling sick and he spent the night at home vomiting.

---

[2] Boris Berezovsky made huge amounts of money in Russia in the 1990s when the country privatised state property. He profited from controlling various assets, including the country's main television channel. After a public disagreement of constitutional reforms made by Vladimir Putin, Berezovsky was granted political asylum in the UK. On 23 March 2013, 7 years after the Litvinenko murder, Berezovsky's bodyguard found him dead in the bathroom of his home with a ligature around his neck. The bathroom door was locked from the inside. The was no evidence of any involvement of chemical, biological or radiological materials. The cause of death was confirmed as hanging and the inquest coroner recorded an open verdict.

FIGURE 12.5   Left: the front of the Millennium Hotel taken from Grosvenor Square. Middle: the hotel main entrance. Right: looking into the Pine Bar from the hotel lobby on the ground floor.

Source: Millennium Hotel. Authors photographs

After Litvinenko had been transferred from Barnet Hospital to University College Hospital on 17 December, he explained his background and described the details of the events to the two police detectives during interviews that took place over the following 3 days. Although Litvinenko was becoming increasingly frail his previous training as a detective in the Russian security services allowed him to give the British detectives a detailed picture of the individuals and the circumstances around his movements on the day of his poisoning. On the 19 November the hospital released information that Litvinenko was deliberately poisoned with thallium and was undergoing further toxicology investigations. On 20 November pictures were released showing his dramatic appearance. He was thin, his hair had fallen out and he looked jaundiced with a yellow skin colour due to loss of liver function resulting in high levels of serum bilirubin, the yellowish brown pigment in bile (Figure 12.6).

On 21 November, there was further uncertainty over what had made Litvinenko ill. Professor John Henry, a toxicologist at St Mary's Hospital London, said Mr Litvinenko may have been poisoned with radioactive

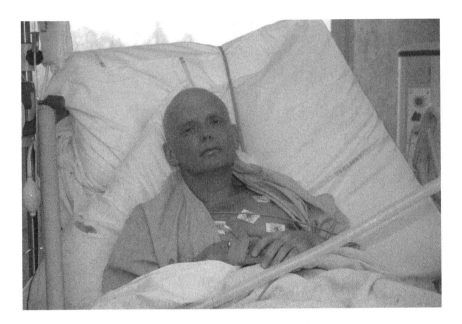

FIGURE 12.6   The photograph of Alexander Litvinenko showing his condition after being admitted to UCH, London.

Source: Litvinenko. www.gettyimages.co.uk/photos/alexander-litvinenko?family=editorial&phrase=alexander%20litvinenko&sort=best#

thallium.[3] The following day Litvinenko's condition deteriorated and he was described as being critically ill. Dr Geoffrey Bellingan, Director of critical care at UCH, ruled out radioactive thallium as the cause of his sickness. On the night of 2 November, Litvinenko had two heart attacks and he died of a further heart attack at 8.51 pm on 23 November. In a statement made by Litvinenko before he died, he accused the Russian President Vladimir Putin of involvement in his death. When the police had first been alerted to possible Russian involvement, they were able to refer to the intelligence records of previous poisonings and potential toxic agents. To identify the nature of Litvinenko's poison, expertise was sought from the UK Ministry of Defence Science and Technology Laboratories at Porton Down in Wiltshire, and the UK Atomic Weapons Establishment at Aldermaston in Berkshire provided further expertise and advice. Following the detailed analysis of urine samples, scientists

---

[3] At that time I was the Honorary Secretary of the British Nuclear Medicine Society. On 21 November, once the news had emerged of the possibility of poisoning with radioactive thallium, the Society's office took a large number of calls from the media enquiring about radioactive thallium. The particular concern was that thallium in the form of thallium-201-thallous chloride was used as a radiopharmaceutical for the investigation of blood flow to the muscle of the heart, an investigation known as a myocardial perfusion study. Having given reassurance of the safety of thallium investigations a week later the offices were overwhelmed with media enquiries and calls from hospital across the world on the nature of the poisoning and the risks concerning polonium-210.

and health experts stated on the day of his death that Litvinenko had been poisoned by the radioactive material polonium-210. This was believed to be an unprecedented act and caused widespread shock and amazement. By 26 November hundreds of people had contacted the National Health Service telephone helpline to seek advice and reassurance about the risks of contamination and radiation poisoning in the London area. Litvinenko's death became a major criminal investigation and as a result of the public concern following the press release the National Health Service telephone enquiry offices received almost 4,000 calls. Management of the situation resulted in an overall cost of around £3 million. The City of Westminster spent £400,000 on environmental health measures and clearing the sites of contamination and the London Metropolitan Police spent nearly £1.5 million on the investigation.

12.4 POLONIUM-210

Polonium-210 is considered to be one of the most toxic naturally occurring radionu-clides found on earth and has the potential to affect human health, due to its wide environmental distribution. Human radiation exposure from polonium results from both ingestion and inhalation. It is a highly toxic radioactive heavy metal with a physical half-life of 138 days and decays to stable lead-206 giving off 5.3 MeV alpha particles that have a range of less than a millimetre in tissue. It occurs naturally in the earth's crust and was the first element to be discovered by Marie and Pierre Curie as they worked to discover the nature of radioactivity in pitchblende ore in 1898. Artificial production normally requires a reactor for the bombardment of bis-muth-209 with neutrons. It has been used in devices that eliminate static and dust, since the alpha particles ionise the air, neutralising static electricity. This has appli-cations for industrial processes such as paper rolling, the production of plastic sheet-ing, spinning synthetic fibres and spray-painting work. Polonium-210 has also been also used in the fibres of brushes to remove dust from photographic film and camera lenses. Large amounts of polonium-210 generate heat as the atoms decay, leading to use in thermoelectric power generators and heaters for a range of military and space applications. The Russian lunar landers used these devices during space missions to warm the operational instruments at night. An ingenious application was considered by the Firestone Tyre and Rubber Company who held a patent for the use polo-nium-210 incorporated in the electrodes of automobile spark plugs (Figure 12.7). The emission of alpha particles was considered to result in a more responsive and reliable spark during engine ignition; however, this was never adopted for commer-cial long-term use.

When used in devices, polonium is normally electroplated onto other metals, making it difficult to separate into a form that can be removed for use as poison. However, polonium-210 is soluble in aqueous solution and forms simple salts (e.g. chloride) in dilute acids.

Prior to the London poisoning there were only a few other recorded events describing the toxic nature of polonium poisoning in humans. The first known case was Nobus Yamada, who died in 1927 after working with polonium in Marie Curie's lab. During the Second World War Dr Robert Fink of the University of

FIGURE 12.7   The Polonium Spark Plug produced by the Firestone Tyre and Rubber
Company.

Source: The Polonium Spark Plug produced by the Firestone Tyre and Rubber Company. Author's
photograph

Rochester gave polonium-210 in water to a patient with myeloid leukaemia and
four others as part of a medical experiment. The cancer patient died, presumably
of his disease and the other four individuals survived. In the years following the
Second World War physicist Dror Sedah, who was working with polonium-210 on
Israel's nuclear programme, reported that when he worked with polonium there
was widespread contamination on everything he touched in his lab and at his
home. This illustrates how difficult it is to contain the material and how easily it
could spread and contaminate items. One of his students subsequently died of leukaemia. There is also a reported case of a Russian male worker who accidentally
inhaled an aerosol estimated to contain approximately 530 MBq of polonium-210.
The total retention was estimated as being approximately 100 MBq, with 13.3
MBq in the lungs, 4.5 MBq in the kidneys and 21 MBq in the liver. At the time
of admission to hospital 2 to 3 days after ingestion, the worker had fever and
severe vomiting, but no diarrhoea. Anyone receiving such amounts would show
symptoms of acute radiation sickness syndrome with bone marrow failure. He
died after 13 days.

Weight for weight, polonium-210 is a million times more toxic than poisons
such as hydrogen cyanide or arsenic. The maximum safe body burden of polo-
nium-210 is only 7 picograms. A microgram, no larger than a speck of dust, could

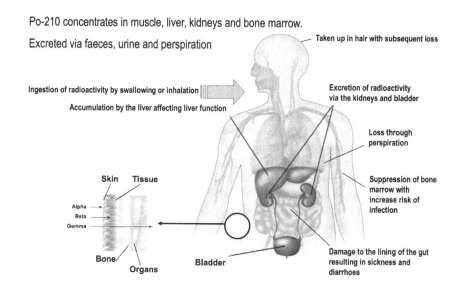

FIGURE 12.8   The biological fate of polonium in the human body.

Source: The biological fate of polonium in the human body. Author's figure

deliver a fatal human dose of radiation. Following ingestion, polonium has a bio-logical half-life of 50 days. It is readily absorbed from the gut into the blood. Once within the bloodstream it is rapidly deposited in major organs and tissues with 30% uptake in the liver, 10% in the kidneys and 5% in the spleen. Around 10% is taken up in bone marrow, leading to disruption of the blood-forming cells and the depletion the blood cells. Figure 12.8 shows the biological pathway and distribu-tion of polonium.

The alpha radiation from the polonium absorbed within the tissues results in a massive destruction of cells, leading to a progressive decline in health. Animal stud-ies have indicated that in an adult male, amounts of polonium-210 absorbed into the blood equal to or greater than 0.1–0.3 GBq would result in death within a month. This would correspond to the ingestion of 1–3 GBq or greater assuming 10% gas-trointestinal absorption to the bloodstream. Within a few hours of intake, remedial medical treatments are considered to be fruitless, since once significant amounts of polonium-210 have entered the bloodstream and deposited in tissues there is nothing that medicine can offer.

12.5 DEPLOYMENT, DIAGNOSIS AND DETECTIVE WORK

The medical team caring for Alexander Litvinenko were initially confused as to the nature of his illness. It took 3 weeks for the doctors to identify the radioactive poison that he had been given. This was due to several reasons, including the clandestine way it was given and because of the physical properties of polonium-210. Radiation monitoring at the bedside had failed to detect evidence of radioactivity, since polonium-210 only produces one gamma ray in every 100,000 decay events. The alpha particles emitted from the radioactive atoms distributed throughout Litvinenko's body were absorbed by his skin and tissues, so nothing was detected externally. It was not until the day of Litvinenko's death that healthcare providers successfully identified the offending poison.

It was known that small amounts of polonium were cleared from the body through urine and faeces. The measurement of polonium levels in urine samples had previously been carried out to monitor workers who were potentially exposed to polonium. The first thing the UK scientists had to determine was to establish the amount of polonium excreted in normal individuals. Human polonium intake occurs from drinking water and everyday foodstuffs. Dietary consumption of lead also adds to the level of polonium in the body, since lead-210 decays to polonium-210. Cigarette smokers, in particular, take in higher amounts of polonium from inhaling tobacco smoke. The biological nature of the tobacco plant makes it an avid absorber of heavy metals from the soil though the roots to the leaves. The presence of high amounts of polonium in tobacco leaves is also due to the soil in the regions where the tobacco plant is grown. The average amount of polonium-210 in a single cigarette is around 17 mBq. The data gathered by the UK health protection team from the previously published studies showed the higher levels of polonium excretion in smokers (Table 12.3).

The scientists working for the UK Health Protection Agency established that the minimum activity of polonium-210 that they could detect in the urine collected from an individual over a 24-hour period was 20 mBq. Accounting for variations in background radiation and the sensitivity of their measuring techniques, they set a reporting threshold of 30 mBq per day. In other words, anyone who produced a 24-hour urine sample that contained above 30 mBq per day would have been exposed to

TABLE 12.3
Reported levels of polonium-210 excretion in urine. Values taken from Bailey et al., HPA Report No. 067. 2010

Country	Activity excreted in urine
USA	9.3 mBq per day (mean)
Germany	2.0–9.9 mBq per day
Brazil	
Non-smokers	3.0–7.4 mBq per day
Smokers	5.8–14.0 mBq pr day
Saudi Arabia	
Non-smokers	1.5–10.0 mBq per litre*
Smokers	3.3–15.9 mBq per litre*
Shisha smokers	2.2–19.9 mBq per litre*

* A healthy adult would normally drink around 2 litres of fluid per day and urinary excretion is around 800 millilitres to 2 litres a day.

polonium-210. The measuring technique involved evaporating one litre of urine to dryness, adding hydrochloric acid and then heating the sample to cause the polonium metal to deposit on a sliver disc. The activity on the disk was then counted for 12 hours on a solid-state alpha spectrophotometer.

The high level of polonium-210 in Alexander Litvinenko's urine confirmed that he had been poisoned with polonium-210. It was subsequently estimated that he had 26.5 milligrams polonium in his bloodstream (less than 1 milligram would be lethal). The post-mortem examination of Litvinenko was carried out by a consultant forensic pathologist 31 days after the poisoning. Because of the large amount of polonium inside his body, the examination was performed using strict safety measures, with a radiation protection officer present inside the room and police guards outside. The post-mortem room was draped in protective sheets to avoid the spread of radioactive contamination and all those attending were wearing protective suits, gloves taped at the wrists and large battery-operated plastic hoods into which filtered air was piped. The main findings were the presence of blood-tinged inflammation of the heart wall (fibrinous pericarditis) with bilateral congestion of the lungs, which contained large amounts of fluid (pleural effusion), fluid in the abdomen (gross ascites) and a generalised break down of the cells in most organs. Polonium-210 was present in all organs and tissues, with the highest activity in the liver (30 MBq/g) and kidneys (49 MBq/g). The lower concentration of polonium in lung tissue (3·5 MBq/g) confirmed that intake of the poison was by swallowing. Assuming the poisoning had taken place 31 days previously and there was 10% absorption to the systemic circulation, the estimated intake was 4,400 MBq (4.4 GBq). The radiation doses to his organs have been estimated to be between 20 and 100 Gy.

By measuring the amount of polonium excreted in urine, the UK health scientists not only showed the high body burden of polonium-210 in Alexander Litvinenko but they also provided a way of checking if anyone who had close contact with him or who had been caring for him had ingested any radioactivity. After screening possible contacts, a total 753 measurements were carried out. Of these, urine samples from 139 individuals contained greater than the reporting level of 30 mBq per day. These individuals were followed up and assessed and reassured as to the associated risk. The highest levels of activity were found in Litvinenko's wife, family and friends, office staff, and patrons and bar staff at the Millennium Hotel. Extensive radiation surveys carried out by the scientific team revealed a trail of contamination across London, at places visited by Litvinenko and his poisoners. The bus Litvinenko took into London on the day of the poisoning showed no signs of radioactivity, but large amounts were detected at the Millennium Hotel. High levels of polonium were found on various objects within the Pine Bar. After testing 100 teapots, one clearly showed high levels of activity despite being in use by other customers for a month with regular cleaning in a dishwasher (Figure 12.9).

Activity was found inside the dishwasher, on a cup, a coffee strainer, a chopping board, an ice cream scoop and on spirit bottles kept behind the bar. Activity was also found on the chairs where the three men sat and on a piano stool. High levels of polonium were also found in the male toilet near the lobby on the ground floor of the hotel, particularly in the second cubicle on the left, the toilet flush handle, in two

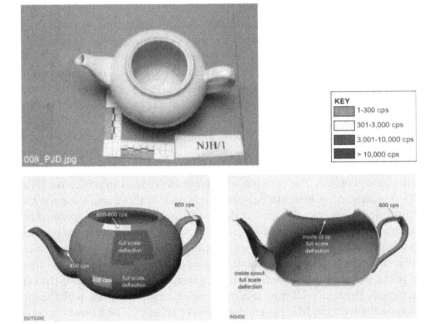

FIGURE 12.9   The teapot found at the Pine Bar that was used to pour the green tea containing polonium-210. The highest level of radioactivity was found in the spout.

Source: Radiation levels marked on the teapot found at the Pine Bar. Figure reproduced from the Litvinenko Inquiry re-used under the UK Open Government Licence

sinks and below the hand dryer. A large amount of radioactive contamination was also found in room 382 of the hotel where Dmitry Kovtun stayed. It would seem that the remainder of the liquid poison was disposed by pouring it down the sink in the room. High readings remained in the accumulated sludge and debris in the sink pipe U bend and also on the room heating control panel and the phone. Polonium was found on the fax machine at Boris Berezovsky's office and the amount of radioactivity left by Litvinenko in the car that took him home to Muswell Hill was so high that the vehicle was considered unsafe to use. Everything that Litvinenko touched at his home during the days after the poisoning was contaminated. Following his death his family was unable to return to the house until 6 months later. In total the Health Protection Agency tested 47 sites across London. Polonium was found at a number of other locations including the Piccadilly Itsu Sushi Bar, The Emirates Football Stadium, Hey Jo Lap dancing club and Abracadabra Bar (which specialised in Russian food), Dar Marrakesh restaurant and a Lambeth-Mercedes taxi. Radioactivity was also found on British Airways aircraft which had been on flights to Hamburg and Moscow.

The extent and location of all the various places where the radioactivity was found was puzzling since they did not fit with the known movements of the three men in November. The detectives solved the clues to the murder when they realised that there were three separate polonium trails in and out of London, which resulted from movements on three different dates. The activity found at the Itsu Suchi Bar was an example of this, since Litvinenko was poisoned after he ate the meal there on 1 November. It was subsequently realised that the contamination had occurred during a previous visit to the restaurant. The investigation concluded that Lugovoy and Kovton had made two previous unsuccessful poisoning attempts to murder Litvinenko before the third and successful assassination. The first attempt took place on 16 October 2006, when radioactive traces were found in all places they visited before and after their meeting with Litvinenko. They had added the poison to his tea, but he did not drink it. A second unsuccessful attempt took place on 25 October, when Lugovoy and Kovtun again flew to London. They left radioactive traces again in their hotel prior to meeting Litvinenko but did not administer the poison, possibly due to the presence of security cameras in the meeting room. There was evidence that they disposed of the poison down the toilet in the hotel room before flying home. The theory that there were three murder attempts was supported by an autoradiograph taken from a strand of Litvinenko's hair. Autoradiographs are obtained by placing a sample of radioactivity on a film or electronic detector. Each alpha particle leaves a dot on the detector allowing an image to build up over a period time. The radiograph clearly showed the polonium activity taken up in Litvinenko's hair as it was growing. Three discrete events can be seen showing the separate intake of the poison as shown in Figure 12.10.

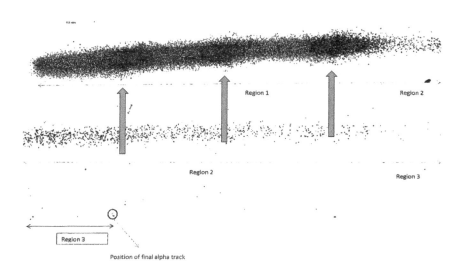

FIGURE 12.10   Autoradiograph of a strand of Alexander Litvinenko's hair. The arrows point to three separate time points where radioactivity was taken up as the hair grew.

Source: Autoradiograph of a strand of Alexander Litvinenko's hair. Figure reproduced from the Litvinenko Inquiry re-used under the UK Open Government Licence

The evidence was given during a UK public enquiry, headed by Sir Robert Owen. In January 2016 the enquiry report concluded that Andrei Lugovoy and Dmitry Kovtun were responsible for the poisoning of Litvinenko. It is thought that this was a state-sponsored action using an exotic poison that could be brought across international boundaries without detection and administered in an unobtrusive manner, but that Lugovoy and Kovtun did not fully realise the nature of what they were handling or of the dangers associated with it. The events have been described in detail in the book *A Very Expensive Poison* written by the *Guardian* journalist Luke Harding. He described the perpetrators actions while handling the radioactive poison in leaky containers as 'idiotic and verging on suicidal'. They seemed to be unaware of the nature of the material they were handling, storing it in their hotel rooms and disposing of the radioactive poison in the toilet and sink, using ordinary towels to clean up leaks and splashes.

Just how someone would get hold of polonium for criminal acts is uncertain. Around 100 grams a year of polonium-210 is manufactured worldwide in nuclear reactors. At the time of the poisoning there were no reported thefts of this material, although there was some concern of lack of security in some eastern European countries. There was one scientist in the UK who was in a position to offer some insight as to the origin of the material. Norman Dombey was Emeritus Professor of theoretical Physics at the University of Sussex in Brighton. He had extensive knowledge of international nuclear weapons programmes, spoke Russian and had spent a year at Moscow state university in the 1960s and visited again in 1988. After various inquiries and experiments he concluded that the polonium had been made at the Mayak nuclear facility in Russia. The very same facility that was described in Chapter 10 and considered to be one of the most radioactive places on earth! Once the material had been produced at the Mayak reactor facility the Russian state poison laboratories had a great deal of knowledge and expertise on how this could be prepared and deployed to commit a political assassination.

12.6 OTHER POLONIUM POISONINGS

In 1957 Nikolai Khokhlov survived an assassination attempt after being given a drink containing what was believed to be radioactive thallium. Although there had been other high-profile political poisonings in the UK (Gorgi Markov, Waterloo Bridge ricin poisoning, and Sergi Skripal, Salisbury Novichok poisoning) the death of Alexander Litvinenko in London on 23 November 2006 was the first proven assassination by the spiking of a drink with a radioactive poison. At the time this was considered to be unprecedented and the surreptitious nature of the murder almost escaped detection by the medical staff and UK authorities. It is difficult to know if the perpetrators ever thought that the true nature of the poison would to be discovered. It is, however, believed that there may have been previous killings of this nature outside the UK. There are at least three unexplained deaths where the victims died with similar symptoms. According to Luke Harding's book *A Very Expensive Poison* one of the victims was the Russian journalist Yuri Shchekochikhin who in the summer of 1988 published an article concerning organised crime in the Soviet Union. In 1995 he was elected to the Russian State Duma and joined the Duma committee dealing with the problems of corruption and organised crime. He went to great efforts to investigate the criminal activities of the Russian security services and raised a series of probing questions. On 3 July 2003 he died suddenly after a mysterious 16-day illness. He had been admitted to the Central Clinical Hospital of the Presidential Administration of the Russian Federation in Moscow, who officially stated that he died from an allergic reaction. His relatives were denied any medical reports on the cause of his illness and they were forbidden to take any tissue specimens for independent medical investigation.

Another suspicious death was that of the Chechen guerrilla commander Lech Ismailov who fought against the Russians in the second Chechen war that took place between 1999 and 2009. After being captured by Russian forces he was considered a valuable source of information on Chechen guerrilla activities. He was given the opportunity of changing sides and work for the Russians security services but refused the offer. In 2005, after being sentenced to 9 years in prison he was being transferred to another prison when he was offered some food and a farewell cup of tea. Soon after his hair fell out and his skin erupted in blisters and he died 9 days later.

The third suspected case is perhaps the most similar to the London poisoning. On 24 September 2004 Roman Igorevich Tsepov died in St Petersburg of poisoning with an unknown material. Tsepov was a St Petersburg businessman with past connections to President Putin. When Putin came to power, Tsepov became one of the most influential figures in the financial and political life of Saint Petersburg, although he had been involved with a number of criminal activities including drugs, illegal weapons and extortion. From 1993, there were five unsuccessful attempts on his life. On 11 September 2004 Tsepov was on a business trip to Moscow and visited colleagues at a Russian FSB security office where he had a cup of tea. Later that day he felt unwell and subsequently developed symptoms of vomiting and diarrhoea. He was treated in Hospital 31 in St. Petersburg, where doctors observed a sudden drop of his white blood cells. Less than 3 weeks later he died. The post-mortem report suggested that radioactive material was used in the murder, with body contamination levels found to be one million times higher than background radiation levels. Tsepov's symptoms prior to death have been described as being similar to severe radiation sickness.

On 6 November 2004, the late President of Palestine, Yasser Arafat died of multiple organ failure one month after the sudden onset of severe nausea, vomiting, abdominal pain and diarrhoea. In 2011 abnormal traces of polonium-210 were found on his belongings and his remains were exhumed in 2012. Swiss researchers from the Institute of Physics at Lausanne University found elevated levels (up to 20 times above normal) of polonium-210 in his bones. At the time of his death there was no evidence of hair loss or of bone marrow suppression. The cause of his death still remains unproven.

Whist the details of these poisonings are difficult to piece together Litvinenko polonium poisoning remains the most thoroughly documented known case. The fact that the nature of polonium poisoning was discovered and the detailed information obtained is a credit to the detectives and scientists involved. There is one sad event concerning one of the radiation scientists who worked on the detection of polonium in the Litvinenko case. This was brought to my attention by one of my colleagues, Professor Phil Blower at Kings College London. One of Prof Blower's past PhD students was Dr Matthew Puncher, who worked in the biomathematics group at the UK Health Protection Agency in Harwell and had carried out the calculations for determining the amount of polonium-210 in Litvinenko's body. As part of an international project funded by the US Department of Homeland Security, Dr Puncher travelled to the Russian Mayak nuclear facility on two occasions to work on software to calculate the amounts of body radiation ingested by workers on the plant. During the second visit he and one of his colleagues felt uncomfortable and thought they were being watched and followed. When Puncher returned home from Russia he was said to have 'changed completely' and was deeply depressed and obsessive about an error he thought he had made in a software program. In May 2016, he was found dead in his home with multiple extensive wounds from two kitchen knives. Home Office pathologist Dr Nicholas Hunt stated that he could not entirely exclude the involvement of someone else, but he considered that the wounds were self-inflicted, causing death by haemorrhaging. There was no evidence of a disturbance or a struggle but committing suicide by self-inflicted knife wounds is extremely rare and the use of two knives was even more concerning.

12.7 SECURITY OF RADIOACTIVE MATERIALS

Prior to the end of the 20th-century radiation safety and support to radiological and nuclear incidents was aimed at treating casualties, containing any contamination and environmental restoration. Planning for the majority of such events was carried out on the premise that this would be of benefit following accidents or in cases human error. Following the 9/11 attacks in New York, the 7/7 London bombings and the terrorist events in France, Australia and New Zealand, emergency services have had to prepare and plan for deliberate nuclear and radiological disruptive and dire actions, such as improvised explosive devices, dirty bombs, the deliberate contamination of water or food and intentional poisoning with radioactivity. In many countries the police, security services and environmental health professionals have been visiting industrial, academic and hospital premises to assess the security of radioactive materials. In hospitals, the emergency care staff in casualty departments are undergoing regular training in how to recognise and treat radiological casualties.

Glossary

Absorbed dose: the amount of energy received per unit mass of tissue from ionising radiation. The SI unit of absorbed dose is gray (Gy)

Absorption: the process by which radiation imparts some or all of its energy to any material through which it passes. In the case of gamma and X-rays this includes processes known as the photoelectric effect, Compton effect and pair production.

Activation: the process through which a stable nuclide is transformed into a radionuclide by being bombarded with particles such as neutrons, protons or alpha particles.

Activity (radioactivity): a measure of the number of atoms disintegrating per second in a sample of material. The unit for activity in Europe is becquerel (Bq) and in the US it is curie (Ci). 1 Bq = 2.703×10^{-11}Ci

Acute exposure: an exposure to radiation that occurred in a matter of minutes rather than in longer, continuing exposure over a period of time.

Acute Radiation Effects: biological effects occurring after a short but very high exposure to ionising radiation.

Acute Radiation Syndrome (ARS): also known as radiation sickness, results when a person is exposed to a high dose of penetrating radiation (e.g. X-rays, gamma rays or neutrons) in a short period of time. There are four subsyndromes (haematopoietic, cutaneous, gastrointestinal and neurovascular) which vary with dose and individual characteristics.

Air burst: A nuclear weapon explosion that is high enough in the air to keep the fireball from touching the ground. The radioactivity in the fallout from an air burst is relatively insignificant compared with a surface burst because the fireball does not reach the ground and does not pick up any surface material.

Alpha particle: a positively charged particle ejected spontaneously during radioactive decay. It is identical to a helium nucleus that has a mass number of 4 and an electric charge of +2.

Annihilation radiation: the electromagnetic radiation resulting from the mutual annihilation of two particles of opposite charge. In the case of a collision between a positron (positively charged electron) and a negative electron, the annihilation radiation consists of two gamma photons each of energy about 0.511 MeV, travelling in opposite directions.

Atom: the smallest particle of an element that can enter into a chemical reaction.

Atomic mass number: the total number of protons and neutrons in the nucleus of an atom.

Atomic number:the total number of protons in the nucleus of an atom.

Atomic weight: also called the relative atomic mass, it is the relative weight of the atom of an element compared with a standard. Since 1961 the standard has been one-twelfth the mass of an atom of carbon-12.

Background radiation: the surrounding ionising radiation from natural sources such as rocks, soil, buildings and cosmic radiation originating in outer space.

Becquerel (Bq): the special name of the unit of activity. One becquerel is equivalent to one atom disintegrating per second. This is a small unit, large multiples are commonly used (e.g. kBq, MBq and GBq).

Beta particles: electrons ejected from the nucleus of a decaying atom.

Biological half-life: the time taken for one half of the amount of a substance, such as a radionuclide, to be expelled from the body by natural metabolic processes, not counting radioactive decay.

Bremsstrahlung: braking radiation. The secondary photon radiation produced by the deceleration of charged particles as they pass through matter. This process includes the production of X-rays when electrons are slowed or stopped by the anode in an X-ray tube.

Cancer incidence: the frequency of occurrence of cases of cancer whether fatal or not. The rate with which new cases of disease occur, for example per year or per million population.

Carcinogenic: a cancer-causing substance.

Chelator: a substance to which atoms may become attached, facilitating their excretion from the body. Also used for linking radiometals to molecules for radiolabelling, for example in radiopharmaceuticals.

Controlled area: a radiation area subject to special rules for the purpose of protection against ionising radiation or of preventing the spread of radioactive contamination. Access is controlled and restricted to individuals who have received appropriate instructions and training.

Critical group of individuals: the group of people identified on grounds of sex, diet, locality, ethnicity, age, etc. who are likely to be those most exposed to radiation from a particular source.

Critical mass: the minimum amount of fissile material that can achieve a self-sustaining nuclear chain reaction.

Criticality: a fission process where the neutron production rate equals the neutron loss rate from absorption or leakage. Criticality occurs when a large enough mass of a fissile material is concentrated within a small enough volume to begin a chain reaction. A nuclear reactor is 'critical' when it is operating.

Curie (Ci): the unit of radioactivity mainly used in the US. One curie of any sample of radioactive material undergoes the same number of nuclear disintegrations per second as occurs in 1 gram of radium-226. $1 \text{ Ci} = 3.7 \times 10^{10}$ Bq.

Cyclotron: a type of particle accelerator invented by Ernest Lawrence that accelerates charged particles outwards for bombarding materials to produce artificial radioactivity. These are commonly used to produce fluorine-18 for PET scanning.

Decay chain: the sequence of radioactive atoms produced by successive transformations from an original unstable radionuclide, terminating when a stable form of atom is finally reached.

Decay products (child or daughter products): the nuclides, elements and particles and electromagnetic radiation produced during radioactive decay.

Disintegration: the spontaneous transformation of the nucleus of a radioactive atom to a new nuclear form with the emission of radiation.

DNA (Deoxyribonucleic acid): self-replicating material that is present in the chromosomes of nearly all living organisms. It is the carrier of genetic information. Four chemical compounds (adenine, cytosine, guanine and thymine) attached in sequence along the DNA molecule determine the coding information of genes.

Dose: the amount of energy absorbed by tissue (see absorbed dose).

Dose limits: the limits laid down by legal authorities for the absorbed doses resulting from ionising radiation to workers, trainees, students and members of the public, excluding the doses resulting from natural background radiation and medical exposures.

Dosimeter: a small portable instrument (such as a film badge, thermoluminescent or pocket dosimeter) for measuring and recording the total accumulated dose of ionising radiation.

Dose rate: the radiation dose delivered per unit time.

Electromagnetic radiation: a traveling wave motion resulting from changing electric or magnetic fields. This includes X-rays and gamma rays that have short wavelengths, through ultraviolet, visible and infrared regions, to radar and radio waves that have relatively long wavelengths.

Electron: the negatively charged particle in all atoms, its positively charged counterpart of equal mass and charge being called the positron. The word electron is often used to include both negative electrons (negatrons) and positive electrons (positrons).

Electron volt (eV): a unit of energy equivalent to the amount of energy gained by an electron when it passes from a point of low potential to a point 1 volt higher in potential.

Element: matter consisting of one type of atom having the same number of protons.

Epidemiology: the study of the frequency of disease or abnormality in human populations.

Excitation: the addition of energy to an atom, transforming it from its ground state to an excited state.

Exposure: the process of being irradiated with or exposed to radiation with the potential to result in an absorbed dose.

Fallout: minute particles of radioactive debris that descend slowly from the atmosphere after a nuclear explosion.

Field: the volume of a beam of radiation such as X-rays or gamma rays in diagnostic or therapeutic radiology.

Film badge: photographic film used for the measurement of personal radiation doses of radiation.

Fission: a nuclear reaction in which a heavy nucleus splits into two (or rarely three or four) approximately equal parts.

Fluorine-18 (F-18): a positron-emitting radionuclide with a half-life of 110 minutes made using a cyclotron or linear accelerator and used for positron emission tomography (PET scanning).

Gamma camera: an instrument that detects gamma rays to produce images of the distribution of a radioactive material within the body.

Gamma rays: high-energy electromagnetic radiation emitted by certain radionuclides. These rays have high energy and a short wave length. All gamma rays emitted from a given radionuclide have the same energy, a characteristic that enables scientists to identify which gamma emitters are present in a sample.

Geiger counter: a radiation detection instrument that consists of a gas-filled tube containing electrodes, between which there is an electrical voltage but no current flowing. When ionising radiation passes through it ionises the gas within the tube producing a short, intense pulse of current that is measured or counted.

Gray (Gy): unit of measurement for absorbed dose. One gray is equal to one joule of energy deposited per kilogram of tissue ($1 \text{ Gy} = \text{J kg}^{-1}$). The unit Gy can be used for any type of radiation but it does not describe the biological effects of the different radiations.

Ground state: the lowest energy level of an atom or system.

Half-life: the time any substance takes to decay by half of its original amount.

Heavy water: water in which hydrogen atoms have been replaced by deuterium, the 'heavy' isotope of hydrogen that has one proton and one neutron giving it twice the mass of ordinary hydrogen. It is mainly used as a coolant in nuclear reactors.

Ion: an electrically charged particle formed when one or more electrons are stripped off an atom.

Ionisation: the process of removing or adding one or more electrons to or from atoms or molecules creating ions.

Intervention: a human activity that decreases the overall exposure of individuals to radiation by removing existing sources, modifying existing exposure pathways or reducing the number of individuals exposed.

Latent period: the time interval (years or decades) between the event causing cellular change with the potential for cancer formation and the development of the detectable disease.

Lethal dose (50/30): the dose of radiation expected to cause death within 30 days to 50% of those exposed without medical treatment. The generally accepted dose is about 4 mSv received over a short period of time.

Lymphocyte: a type of white cell in blood that is part of the immune system. There are two main types: T cells and B cells. B cells produce antibody molecules that can latch on and destroy invading viruses or bacteria.

Magnox: a type of nuclear reactor in which the fuel rods are sheathed in magnesium oxide. These have been in operation in the UK since 1956.

Medical exposure: the intentional irradiation of a person, either externally or internally, for the purpose of his or her own medical treatment or diagnosis or as the subject of medical research.

Metabolite: material formed as a result of metabolic processes (metabolism) in the body.

Metric prefix: a unit prefix that precedes a unit of measure to indicate a multiple or submultiple of the unit. For example microsievert (μSv) is one millionth of a sievert, millisievert (mSv) is one thousandth of a sievert and kBq is a thousand becquerels and megabecquerel (MBq) is one million becquerels. The numerical factors are shown in Appendix 1.

Moderator: materials such as graphite or heavy water between the fuel rods in the core of a nuclear reactor used to slow down and reduce the energy of neutrons to increase the likelihood of further nuclear fission reactions.

Molecule: a combination of two or more atoms that are chemically bonded. A molecule is the smallest unit of a compound that can exist by itself and retain all of its chemical properties.

Neutron: a small atomic particle possessing no electrical charge typically found within the nucleus of an atom.

Nucleus: the positively charged central portion of an atom which is associated with most of the mass but only a minute part of its volume.

Nuclide: a general term applicable to all atomic forms of an element. Nuclides are characterised by the number of protons and neutrons in the nucleus, as well as by the amount of energy contained within the atom.

Organ: a configuration of body cells of a certain type occurring together in the same structure, for example in liver, lung or brain.

Phantom: a model or simulation of the human body constructed of materials allowing the same transmission and absorption of radiation as in the body, so that measurements can be made to estimate the doses received at different positions.

Photon: a single 'packet' or quantum of electromagnetic radiation. Photons have no mass and travel at the speed of light.

Pitchblende: a brown to black mineral that has a distinctive lustre. It consists mainly of uraninite (UO_2) and also contains radium (Ra).

Platelet: the blood cell also known as a thrombocyte that reacts to bleeding forming blood clots.

Plutonium (Pu): a heavy, man-made, radioactive metallic element. The most important is Pu-239, which has a half-life of 24,000 years and can be used in reactor fuel and nuclear weapons.

Polonium (Po): a rare and highly radioactive metal with no stable isotopes. Polonium has the atomic number 84 and is chemically similar to selenium and tellurium. Its natural occurrence is limited to tiny traces of polonium-210 with a half-life of 138 days found in uranium ores.

Positron: a positively charged electron.

Positron emission tomography (PET): the scanning technique of obtaining reconstructed images of the distribution of positron-emitting radiopharmaceuticals such as flourine-18. This requires a scanner comprising two opposing detectors for the detection of 511 keV gamma rays which are emitted travelling in opposite direction.

Proton: a small atomic particle with a positive electrical charge, typically found within the nucleus of an atom.

Radioactive concentration: also known as the radiochemical concentration and is the total amount of radioactivity per unit volume. Usually measured in Bq per litre or Bq per millilitre.

Radioactive contamination: the spread of a radioactive material from a source to any surface, person or environment. This can be in the form of a solid, liquid or gas. Contamination of the human body includes both external skin contamination and internal contamination.

Radioactive decay: spontaneous transformation of the nucleus of a radioactive atom to a new nuclear form with the emission of radiation

Radioactivity: the property of certain nuclides that emit radiation by the spontaneous transformation of their nuclei.

Radioisotope: an isotope which is radioactive.

Radionuclide: a nuclide which is radioactive.

Radiopharmaceutical: a radiolabelled molecule or compound administered for the purpose of medical diagnosis or therapy.

Radium (Ra): a naturally occurring radioactive metal. Radium is a radionuclide formed by the decay of uranium and thorium in the environment.

Scintillation detector: a radiation detector that uses a solid or liquid scintillator and that has the property of luminescence. A light flash (scintillation) is produced when ionising radiation such as a gamma ray is absorbed by the scintillator, allowing detection of the event using a photocell.

Sealed source: radioactive material normally in the form of solid pellets or a powder contained within a sealed canister.

Shielding: the material between a radiation source and a potentially exposed person that can reduce exposure.

SI units: the Systeme Internationale (or International System) of units and measurements. This system of units officially came into being in October 1960 and has been adopted by nearly all countries, although the amount of actual usage varies considerably.

Sievert (Sv): the unit of effective dose to express the health risk of doses ionising radiation. 1 Sv = 1 Joule per kilogram.

Technetium-99m (Tc-99m): a radioactive metallic, metastable element with atomic number 43 and atomic mass 99 that decays emitting 140 keV gamma rays. The first synthetic element which is used extensively in nuclear medicine imaging.

Track: the route of a particle of any form of ionising radiation through tissue or any substance.

Tritium (H-3): a radionuclide of hydrogen.

Unstable nucleus: a nucleus that contains an uneven number of protons and neutrons and seeks to reach equilibrium through the process of radioactive decay.

Uranium (U): a naturally occurring radioactive element containing uranium-238 (U-238) and uranium-235 (U-235).

Whole body exposure: an exposure of radiation to the entire body rather than an isolated part.

X-ray: electromagnetic radiation caused by deflection of electrons from their original paths or inner orbital electrons that change their orbital levels around the atomic nucleus. X-rays, like gamma rays, can travel long distances through air and most other materials.

Bibliography

CHAPTER 1

Aldersey-Williams H. Periodic tales. London: Penguin Books; 2012.

Chao JH, Tseng CL, Hsieh WA, Hung DZ and Chang WP. Dose estimation for repeated phosphorus-32 ingestion in human subjects. Appl Radiat Isot. 2001;54:123–9.

Johnson R. Database of radiological Incidents and related events. 2019. Available from: www.johnstonsarchive.net/nuclear/radevents/index.html

Khokhlov N. In the name of conscience. London: Frederick Muller Ltd; 1960.

Mohtadi H, Murshid A. A global chronology of incidents of chemical, biological, radioactive and nuclear attacks: 1950–2005. National centre for food protection and defence; 2006 July. Available from: https://cpb-us-w2.wpmucdn.com/sites.uwm.edu/dist/0/252/files/2016/07/A-Global-Chronology-of-Incidents-of-Chemical-Biological-and-Radionuclear-Attacks.doc-1u8sbvu.pdf

What is a poison? Note produced by a working party of the environment, health and safety committee [EHSC] of the royal society of chemistry; 2015 Jan. Available from: https://edu.rsc.org/download?ac=15891

CHAPTER 2

Chang I. The rape of Nanking: the forgotten holocaust of world war II. London: Penguin Books; 1997.

Unethical human experimentation in the United States. Available from: https://en.wikipedia.org/wiki/Unethical_human_experimentation_in_the_United_States

Unit 731. Available from: https://en.wikipedia.org/wiki/Unit_731

Voldarsky B. The KGB's poison factory. London: Zenith Press, Frontline Books; 2009.

World Medical Association declaration of Helsinki. Bulletin of the World Health Organization. 2001;79:373–4. Available from: www.who.int/bulletin/archives/79(4)373.pdf

CHAPTER 3

Broadbent MV, Hubbard LB. Science and perception of radiation risk. Radiographics. 1992;12:381–92.

Clarke RH, Valentin J. The history of ICRP and the evolution of its policies. ICRP Publication 109. Ann ICRP. 2009;39:75–110.

Feldman A. A sketch of the technical history of radiology from 1896 to 1920. Radiographics. 1989;9:1113–18.

Kassabian MK. Röntgen rays and electro-therapeutics: with chapters on radium and phototherapy. Packard FR, editor. Philadelphia: J B Lippincott Company; 1910. Available from: https://archive.org/details/rntgenrayselectr00kass/page/n9/mode/2up

Posner E. Reception of Roentgen's discovery in Britain. BMJ. 1970;4:357–60.

Sansare K, Khanna V, Karjodkar F. Early victims of X-rays: a tribute and current perception. Dentomaxillofac Radiol. 2011;40:123–5.

Ulrich H. Incidence of leukaemia in radiologists. N Engl J Med. 1946 Jan 10;234:45–6.

Wojcik A, Harms-Ringdahl M. Radiation protection biology then and now. Int J Radiat Biol. 2019;95(7):841–50. DOI:10.1080/09553002.2019.1589027.

CHAPTER 4

"X" rays as depilatory. Lancet, 1896;i:1296.

Daniel J. Depilatory action of x-rays. Med Records. 1896;49:595–6.

Herzig R. Removing roots: "North American Hiroshima Maidens" and the x ray. Technol Cult. 1999;40:723–5.

Lapidus SM. The tricho system: hypertrichosis, radiation, and cancer. J Surg Oncol. 1976;8:267–74.

Lewis L, Caplan PE. The shoe-fitting fluoroscope as a radiation hazard. California Med. 1950;72:26–30. Available from: www.ncbi.nlm.nih.gov/pmc/articles/PMC1520288/

Shore RE, et al. Skin cancer after X-ray treatment for scalp ringworm. Radiat Res. 2002;157:410–18.

Shvarts S, Romem P, Romem Y, Shani M. The mass campaign to eradicate ringworm among the Jewish community in Eastern Europe, 1921–1938. Am J Public Health Res. 2013;103:e56–e66.

CHAPTER 5

Course BM. A reflection on the 150 anniversary of the birth of Marie Curie. Appl Radiat Isot. 2017;130:280–4.

Humm JL, Sartor O, Parker C, Bruland OS, Macklis R. Radium-223 in the treatment of osteoblastic metastases: a critical clinical review. Int J Radiat Oncol Biol Phys. 2015;91:898–906.

Kovarik B, Neuzil M. Radium girls. Environmental history timeline. Available from: http://environmentalhistory.org/people/radiumgirls/

Moore K. The radium girls. London: Simon and Shuster; 2016. ISBN:978-1-4711-5388-4

Mould RF. The early years of radiotherapy with emphasis on X-ray and radium apparatus of radiology. BJR. 1995;68:567–82. DOI:10.1259/0007-1285-68-810-567

Mould RF. Radium history mosaic. Nowotwory J Oncol. 2007;57(4). PL ISSN:0029-540X

Mullner R. Deadly glow: the radium dial worker tragedy. Washington, DC: American Public Health Association; 1999. ISBN:13-978-0875532455

Oak Ridge Associated Universities. Health physics historical instrumentation collection. Available from: www.orau.org/ptp/museumdirectory.htm

Rowland RE. Radium in humans: a review of US studies. Argonne, IL: Argonne National Laboratory, US Printing Office; 1994.

CHAPTER 6

Brown K. Plutopia. New York: Oxford University Press; 2013.

The Cecil Kelley criticality accident. The origin of the Los Alamos human tissue analysis program. Los Alamos Sci. 1995;23:250–1. Available from: https://permalink.lanl.gov/object/tr?what=info:lanl-repo/lareport/LA-UR-95-4005-13

Cooper NG, editor. Radiation protection and the human radiation experiments. Los Alamos Sci. 1995;23. Available from: https://fas.org/sgp/othergov/doe/lanl/pubs/00326644.pdf

Hempleman LH, Lushbaugh CC, Voelz G. What happened to the survivors of the early Los Alamos nuclear accidents? Conference for Radiation Accident Preparedness. Oak Ridge, TN; 1979, Oct 19–20. Available from: www.orau.org/ptp/pdf/accidentsurvivorslanl.pdf

McLaughlin TP, Monahan SP, Pruvost NL, Frolov VV, Ryazanov BG, Sviridov V. A review of criticality accidents Los Alamos national laboratory. 2000 revision. LA-13638.74-75. Available from: www.orau.org/ptp/Library/accidents/la-13638.pdf

Pais A. J Robert Oppenheimer: a life. New York: Oxford University Press; 2006.

Ponchin E. Nuclear radiation: risks and benefits. Oxford: Clarence Press; 1983.

CHAPTER 7

The advisory committee on human radiation experiments. Final Report. Washington, DC: US Government Printing Office; 1995 Oct. Available from: https://fowlchicago.files.wordpress.com/2014/02/advisorycommitte00unit.pdf

Cooper NG, editor. Radiation protection and the human radiation experiments. Los Alamos Sci. 1995;23.

DOE openness: human radiation experiments. Available from: https://ehss.energy.gov/ohre/index.html

The Redfern Inquiry into human tissue analysis in UK nuclear facilities – Vol 2. Summary London Stationary Office; 2010. ISBN:9780102966183, HC 571 2010-11. Available from: https://webarchive.nationalarchives.gov.uk/20101210152913/www.theredferninquiry.co.uk/public-statements/final-report-redfern-inquiry

Rolston N. Human plutonium injection experiments. Stanford University; 2015. Available from: http://large.stanford.edu/courses/2015/ph241/rolston2/

Standring WJF, Dowdall M, Strand P. Overview of dose assessment developments and the health of riverside residents close to the "Mayak" PA facilities, Russia. Int J Environ Res Public Health. 2009;6:174–99.

Trials of war criminals: before the Nuremberg military tribunals under control council law no. 10. Vol. I. Washington, DC: U.S. Government Printing Office, 1949, pp. 49–50, 720, 980. Available from: https://www.loc.gov/rr/frd/Military_Law/NTs_war-criminals.html

Welsome E. The plutonium files: America's secret medical experiments in the cold war. New York: The Dial Press; 1999.

World Medical Association declaration of Helsinki: ethical principles for medical research involving human subjects. World Medical Association; 2008. Available from: www.wma.net/en/30publications/10policies/b3/index.html

CHAPTER 8

Beyer T, Townsend DW, Brun T, Kinahan PE, Charron M, Roddy R, et al. A combined PET/CT scanner for clinical oncology. J Nucl Med. 2000;41:1369–79.

Herbert R, Kulke W, Shepherd RT. The use of technetium 99m as a clinical tracer element. Postgrad Med J. 1995;41:656–62. doi:10.1136/pgmj.41.481.656. PMC:2483197. PMID:5840856.

Hevesy G. Adventures in radioisotope research. Vol I. New York: Pergamon Press; 1962.

Kroeger EA, Rupp A, Gregor J. Misuse of a medical radioisotope: [125]I labelled playing cards in Germany, a case study. Health Phys. 2020;119:128–32.

Myers WG. Georg Charles de Hevesy: the father of nuclear medicine. J Nucl Med Technol. 1996;24:291–4.

CHAPTER 9

Akashi M, Hirama T, Tanosaki S, Kuroiwa N, Nakagawa K, Tsuji H, et al. Initial symptoms of acute radiation syndrome in the JCO criticality accident in Tokai-mura. J Radiat Res. 2001;42(Suppl):S157–66. DOI:10.1269/jrr.42.s157

Bavestock K, Williams D. The chernobyl accident 20 years on: an assessment of the health consequences and the international response. Environ Health Perspect. 2006;114:1312–17.

Chernobyl: Assessment of radiological and health impacts (2002). Update of chernobyl: ten years on. Paris: OECD; 2002. ISBN:92-64-18487-2

Consequences of the Chernobyl accident and their remediation: twenty years of experience. Report of the Chernobyl forum expert group 'environment'. Vieanna: IAEA; 2006. Available from: https://www-pub.iaea.org/mtcd/publications/pdf/pub1239_web.pdf

Harada M, translator. NHK-TV "Tokaimura criticality accident" crew. A slow death: 83 days of radiation sickness. New York: Vertical Inc; 2015. ISBN:978-1-942993-54-4

World nuclear association information library. Chernobyl accident; 1986 (cited 2002 Apr). Available from: www.world-nuclear.org/information-library/safety-and-security/safety-of-plants/chernobyl-accident.aspx

CHAPTER 10

Dobrzyński L, Fornalski KW, Feinendegen LE. Cancer mortality among people living in areas with various levels of natural background radiation. Dose Response. 2015 Jul–Sep;13(3):1559325815592391.

Seo S, Ha W-H, Kang J-K, Lee D, Park S, Kwon T-E, et al. Health effects of exposure to radon: implications of the radon bed mattress incident in Korea. Epidemiol Health. 2019;41:e2019004. DOI:10.4178/epih.e2019004

Silverstein K. The radioactive boy scout. London and New York: Fourth Estate; 2004. ISBN:1-8415-229-3

CHAPTER 11

Koenig KL, Goans RE, Hatchett RJ, Mettler FA, Schumacher TA, Noji EK, MD, et al. Medical treatment of radiological casualties: current concepts. Ann Emerg Med. 2005;45:643–65.

Lim MK. Cosmic rays: are air crew at risk? Occup Environ Med. 2002;59:428–32.

Manobe M, Ando K. Drinking beer reduces radiation-induced chromosome aberrations in human lymphocytes. J Radiat Res. 2002;43:237–45.

Martin CJ, Harrison JD, Rehani MM. Effective dose from radiation exposure in medicine: past, present and future. Physica Medica. 2020;79:87–92.

Xin N, Li YJ, Li X, Wang X, Li Y, Zhang X, et al. Dragon's blood may have radioprotective effects in radiation induced rat brain injury. Radiat Res. 2012;178:75–85.

CHAPTER 12

Bailey MR, Birchall A, Etherington G, Fraser G, Wilkins BT, Bessa Y, et al. Individual monitoring conducted by health protection agency in the London polonium-210 incident. HPA-RPD-067. Chilton: Health Protection Agency; 2010. ISBN:978-0-85951-666-2. Available from: https://assets.publishing.service.gov.uk/government/uploads/system/uploads/attachment_data/file/337140/HPA-RPD-067_for_website.pdf; www.hpa.org.uk/polonium/default.htm

Froidevaux P, Bochud F, Baechler S, Castella V, Augsburger M, Bailat M, Michaud K, Straub M, Pecchia M, Jenk TM, Uldin T, Mangin P. (210)Po poisoning as possible cause of death: forensic investigations and toxicological analysis of the remains of Yasser Arafat. Forensic Sci Int. 2016;259:1–9.

Goldfarb A, Litvinenko M. Death of a dissident. London: Simon and Schuster; 2007. ISBN:978-1-84737-091-4

Harding L. A very expensive poison. London: Guardian Books; 2016. ISBN:978-1-78335-0933

Harrison J, Leggett R, Lloyd D, Phipps A, Scott B. Polonium-210 as a poison. J Radiol Protection. 2007;27:17–40.

Health Protection Agency. Human radiosensitivity. Report of the independent advisory group on ionising radiation. Documents of the health protection agency; 2013 Mar. ISBN:978-0-85951-740-9.

Kaplan K, Maugh TH. Polonium-210's quiet trail of death. Available from: www.mjwcorp.com/rad_dose_assessments_poloniumarticle.php

Macrakis K. Seduced by secrets. In: Inside the Stasi's spy-tech world. New York: Cambridge University Press; 2008. eBook ISBN:13-978-0-511-38645-9, Hardback ISBN:13-978-0-521-88747-2

Mohtadi H, Murshid A. A global chronology of incidents of chemical, biological, radioactive and nuclear attacks: 1950–2005. National Centre for Food Protection and Defence. 2006 Jul 7. Available from: https://cpb-us-w2.wpmucdn.com/sites.uwm.edu/dist/0/252/files/2016/07/A-Global-Chronology-of-Incidents-of-Chemical-Biological-and-Radio-nuclear-Attacks.doc-1u8sbvu.pdf

Nathwani AC, Down JF, Goldstone J, Yassin J, Dargan PI, Virchis A, et al. Polonium-210 poisoning: a first-hand account. Lancet. 2016;388:1075–80.

Nuclear Forensics Support. IAEA nuclear security series no. 2. Vienna: IAEA; 2006.

Turai I, Veress K, Gunalp B, Souchkevitch G. Medical response to radiation incidents and radionuclear threats. BMJ. 2004;328:568–72.

Tyhorn FG, Widdowson EE. Identification of objects by radioactive labelling. Police J: Theo Pract Princ. 1940;13:45–52. DOI:10.1177/0032258X4001300105

Vogel H, Lotz P, Vogel B. Ionizing radiation in secret services' conspirative actions. Eur J Radiol. 2007;63:263–9.

Appendix 1
Table of SI Metric Prefixes

Prefix	Symbol	Multiplier	Exponential
yotta	Y	1,000,000,000,000,000,000,000,000	10^{24}
zetta	Z	1,000,000,000,000,000,000,000	10^{21}
exa	E	1,000,000,000,000,000,000	10^{18}
peta	P	1,000,000,000,000,000	10^{15}
tera	T	1,000,000,000,000	10^{12}
giga	G	1,000,000,000	10^{9}
mega	M	1,000,000	10^{6}
kilo	k	1,000	10^{3}
hecto	h	100	10^{2}
deca	da	10	10^{1}
		1	10^{0}
deci	d	0.1	10^{-1}
centi	c	0.01	10^{-2}
milli	m	0.001	10^{-3}
micro	µ	0.000001	10^{-6}
nano	n	0.000000001	10^{-9}
pico	p	0.000000000001	10^{-12}
femto	f	0.000000000000001	10^{-15}
atto	a	0.000000000000000001	10^{-18}
zepto	z	0.000000000000000000001	10^{-21}
yocto	y	0.000000000000000000000001	10^{-24}

Index

Note: Page numbers in *italics* indicate a figure and page numbers in **bold** indicate a table on the corresponding page. Page numbers followed by 'n' indicate a note.

Printed in the United States
by Baker & Taylor Publisher Services